AF321608

The Big Data Agenda: Data Ethics and Critical Data Studies

Annika Richterich

University of Westminster Press
www.uwestminsterpress.co.uk

Published by
University of Westminster Press
115 Cavendish Street
London W1W 6XH
www.uwestminsterpress.co.uk

Text ©Annika Richterich 2018

First published 2018

Series cover concept: Mina Bach (minabach.co.uk)

Printed and bound by CPI Group (UK) Ltd, Croydon, CR0 4YY
Print and digital versions typeset by Siliconchips Services Ltd.

ISBN (Hardback) 978-1-911534-72-3
ISBN (Paperback) 978-1-911534-97-6
ISBN (PDF): 978-1-911534-73-0
ISBN (EPUB): 978-1-911534-74-7
ISBN (Kindle): 978-1-911534-75-4

DOI: https://doi.org/10.16997/book14

This work is licensed under the Creative Commons Attribution-NonCommercial-NoDerivatives 4.0 International License. To view a copy of this license, visit http://creativecommons.org/licenses/by-nc-nd/4.0/ or send a letter to Creative Commons, 444 Castro Street, Suite 900, Mountain View, California, 94041, USA. This license allows for copying and distributing the work, providing author attribution is clearly stated, that you are not using the material for commercial purposes, and that modified versions are not distributed.

The full text of this book has been peer-reviewed to ensure high academic standards. For full review policies, see: http://www.uwestminsterpress.co.uk/site/publish/

An electronic version of this book is freely available, thanks to the support of libraries working with Knowledge Unlatched. KU is a collaborative initiative designed to make high quality books Open Access for the public good. More information about the initiative and details about KU's Open Access programme can be found at www.knowledgeunlatched.org

Suggested citation:
Richterich, Annika. 2018. *The Big Data Agenda: Data Ethics and Critical Data Studies.* London: University of Westminster Press. DOI: https://doi.org/10.16997/book14. License: CC-BY-NC-ND 4.0

To read the free, open access version of this book online, visit https://doi.org/10.16997/book14 or scan this QR code with your mobile device: 

Acknowledgments

While working on this book, I have immensely benefitted from the fantastic support of many peers, colleagues and friends. I am very grateful for their advice and encouragement. I would also like to thank those who have contributed to big data discourses, ethics and critical data studies: their insights and their initiation of much-needed debates were invaluable for my work.

I am very grateful to my colleagues at Maastricht University's Faculty of Arts & Social Science. Although more colleagues deserve gratitude, I would particularly like to thank Sally Wyatt, Anna Harris, Vivian van Saaze, Tamar Sharon, Tsjalling Swierstra and Karin Wenz. Their work, advice and support were crucial for this project. My sincere thanks go to Sally Wyatt for her advice and for endorsing my application for a Brocher fellowship. This 1-month visiting fellowship at the Brocher Foundation (www.brocher.ch) allowed me to focus on my book and I would like to thank the foundation as well as its staff. In addition, I tremendously appreciated and enjoyed the company of and the discussions with the other fellows; among them were Laura Bothwell, Alain Giami, Adam Henschke, Katherine Weatherford Darling, Bertrand Taithe, Peter West-Oram and Sabine Wildevuur.

I received detailed, much-appreciated feedback and suggestions from the anonymous reviewers. I would like to thank them for their time and their elaborate comments which were incredibly helpful for revising the manuscript. Moreover, I am very grateful to Andrew Lockett, from University of

Westminster Press, and Christian Fuchs, editor of the *Critical, Digital and Social Media Studies* series, for supporting this book and for enabling its open access publication.

Last, though certainly not least, I would like to thank my family, my parents and my brothers, for being as supportive and understanding as ever. I would like to thank my partner, Stefan Meuleman, not only for patiently allowing me the time and space needed to complete this project, but also for his advice on the book and for making sure that I take time away from it too.

Competing Interests

The author declares that she has no competing interests in publishing this book.

Contents

CHAPTER 1

Introduction

In times of big data and datafication, we should refrain from using the term 'sharing' too lightly. While users want, or need, to communicate online with their family, friends or colleagues, they may not intend their data to be collected, documented, processed and interpreted, let alone traded. Nevertheless, retrieving and interrelating a wide range of digital data points, from, for instance. social networking sites, has become a common strategy for making assumptions about users' behaviour and interests. Multinational technology and internet corporations are at the forefront of these datafication processes. They control, to a large extent, what data are collected about users who embed various digital, commercial platforms into their daily lives.

Tech and internet corporations determine who receives access to the vast digital data sets generated on their platforms, commonly called 'big data'. They define how these data are fed back into algorithms crucial to the content that users subsequently get to see online. Such content ranges from advertising to information posted by peers. This corporate control over data has given rise to considerable business euphoria. At the same time, the power exercised with data has increasingly been the subject of bewilderment, controversies, concern and activism during recent years. It has been questioned at whose cost the Silicon Valley mantra 'Data is the new oil'[1] is being put into practice. It is questioned whether this view on data is indeed such an alluring prospect for societies relying increasingly on digital technology, and for individuals exposed to datafication.

Datafication refers to the quantification of social interactions and their transformation into digital data. It has advanced to an ideologically infused '[...] leading principle, not just amongst technoadepts, but also amongst scholars who see datafication as a revolutionary research opportunity to investigate human conduct' (van Dijk 2014, 198). Datafication points to the widespread ideology of big data's desirability and unquestioned superiority, a tendency termed 'dataism'

How to cite this book chapter:
Richterich, A. 2018. *The Big Data Agenda: Data Ethics and Critical Data Studies.*
    Pp. 1–14. London: University of Westminster Press. DOI: https://doi.
    org/10.16997/book14.a. License: CC-BY-NC-ND 4.0

by van Dijk (2014). This book starts from the observation that datafication has left its mark not only on corporate practices, but also on approaches to scientific research. I argue that, as commercial data collection and research become increasingly entangled, interdependencies are emerging which have a bearing on the norms and values relevant to scientific knowledge production.

Big data have not only triggered the emergence of new research approaches and practices, but have also nudged normative changes and sparked controversies regarding how research is ethically justified and conceptualised. Big data and datafication 'drive' research ethics in multiple ways. Those who deem the use of big data morally reasonable have normatively framed and justified their approaches. Those who perceive the use of big data in research as irreconcilable with ethical principles have disputed emerging approaches on normative grounds. What we are currently witnessing is a coexistence of research involving big data and contested data ethics relevant to this field. I explore to what extent these positions unfold in dialogue with (or in isolation from) each other and relevant stakeholders.

This book interrogates entanglements between corporate big data practices, research approaches and ethics: a domain which is symptomatic of broader challenges related to data, power and (in-)justice. These challenges, and the urgent need to reflect on, rethink and recapture the power related to vast and continually growing 'big data' sets have been forcefully stressed in the field of critical data studies (Iliadis and Russo 2016; Dalton, Taylor and Thatcher 2016; Lupton 2015; Kitchin and Lauriault 2014; Dalton and Thatcher 2014). Approaches in this interdisciplinary research field examine practices of digital data collection, utilisation, and meaning-making in corporate, governmental, institutional, academic, and civic contexts.

Research in critical data studies (CDS) deals with the societal embeddedness and constructedness of data. It examines significant economic, political, ethical, and legal issues, as well as matters of social justice concerning data (Taylor 2017; Dencik, Hintz and Cable 2016). While most companies have come to see, use and promote data as a major economic asset, allegedly comparable to oil, CDS emphasises that data are not a mere commodity (see also Thorp 2012). Instead, many types of digital data are matters of civic rights, personal autonomy and dignity. These data may emerge, for example, from individuals' use of social networking sites, their search engine queries or interaction with computational devices. CDS researchers analyse and examine the implications, biases, risks and inequalities, as well as the counter-potential, of such (big) data. In this context, the need for qualitative, empirical approaches to data subjects' daily lives and data practices (Lupton 2016; Metcalf and Crawford 2016) has been increasingly stressed. Such critical work is evolving in parallel with the spreading ideology of datafication's unquestioned superiority: a tendency which is also noticeable in scientific research.

Many scientists have been intrigued by the methodological opportunities opened up by big data (Paul and Dredze 2017; Young, Yu and Wang 2017; Paul et al. 2016; Ireland et al. 2015; Kramer, Guillory and Hancock 2014; Chunara et al. 2013; see also Chapter 5). They have articulated high hopes about the contributions big data could make to scientific endeavours and policy making (Kettl 2017; Salganik 2017; Mayer-Schönberger and Cukier 2013). As I show in this book, data produced and stored in corporate contexts increasingly play a part in scientific research, conducted also by scholars employed at or affiliated with universities. Such data were originally collected and enabled by internet and tech companies owning social networking sites, microblogging services and search engines.

I focus on developments in public health research and surveillance, with specific regard to the ethics of using big data in these fields. This domain has been chosen because data used in this context are highly sensitive. They allow, for example, for insights into individuals' state of health, as well as health-relevant (risk) behaviour. In big data-driven research, the data often stem from commercial platforms, raising ethical questions concerning users' awareness, informed consent, privacy and autonomy (see also Parry and Greenhough 2018, 107–154). At the same time, research in this field has mobilised the argument that big data will make an important contribution to the common good by ultimately improving public health. This is a particularly relevant research field from a CDS perspective, as it is an arena of promises, contradictions and contestation. It facilitates insights into how technological and methodological developments are deeply embedded in and shaped by normative moral discourses.

This study follows up earlier critical work which emphasises that academic research and corporate data sources, as well as tools, are increasingly intertwined (see e.g. Sharon 2016; Harris, Kelly and Wyatt 2016; Van Dijck 2014). As Van Dijck observes, the commercial utilisation of big data has been accompanied by a '[...] gradual normalization of datafication as a new paradigm in science and society' (2014, 198). The author argues that, since researchers have a significant impact on the establishment of social trust (206), academic utilisations of big data also give credibility to their collection in commercial contexts the societal acceptance of big data practices more generally.

This book specifically sheds light on how big data-driven *public health research* has been communicated, justified and institutionally embedded. I examine interdependencies between such research and the data, infrastructures and analytics shaped by multinational internet/tech corporations. The following questions, whose theoretical foundation is detailed in Chapter 2, are crucial for this endeavour: What are the broader *discursive conditions* for big data-driven health research: Who is affected and involved, and how are certain views fostered or discouraged? Which *ethical arguments* have been discussed: How is big data research ethically presented, for example as a relevant, morally right, and societally valuable way to gain scientific insights into public health? What *normativities* are at play in presenting and (potentially) debating big data-driven research on public health surveillance?

I thus emphasise two analytical angles: first, the discursive conditions and power relations influencing and emerging in interaction with big data research; second, the values and moral arguments which have been raised (e.g. in papers, projects descriptions and debates) as well as implicitly articulated in research practices. I highlight that big data research is inherently a ground of normative framing and debate, although this is rarely foregrounded in big data-driven health studies. To investigate the abovementioned issues, I draw on a pragmatist approach to ethics (Keulartz et al. 2004). Special emphasis is placed on Jürgen Habermas' notion of 'discourse ethics' (2001 [1993], 1990). This theory was in turn inspired by Karl-Otto Apel (1984) and American pragmatism. It will be introduced in more detail in Chapter 2.

Already at this point it is important to stress that the term 'ethical' in this context serves as a qualifier for the *kind* of debate at hand – and not as a normative assessment of content. Within a pragmatist framework, something is ethical because values and morals are being negotiated. this means that 'unethical' is not used to disqualify an argument normatively. Instead, it would merely indicate a certain quality of the debate, i.e. that it is not dedicated to norms, values, or moral matters. A moral or immoral decision would be in either case an ethical issue, and '[w]e perform ethics when we put up moral routines for discussion' (Swierstra and Rip 2007, 6).

To further elaborate the perspective taken in this book, the following sections expand on key terms relevant to my analysis: *big data* and *critical data studies*. Subsequently, I sketch main objectives of this book and provide an overview of its six chapters.

Big Data: Notorious but Thriving

In 2018, the benefits and pitfalls of digital data analytics were still largely attributed to a concept which had already become somewhat notorious by then: big data. This vague umbrella term refers to the vast amounts of digital data which are being produced in technologically and algorithmically mediated practices. Such data can be retrieved from various digital-material social activities, ranging from social media use to participation in genomics projects.[2]

Data and their analysis have of course long been a core concern for quantitative social sciences, the natural sciences, and computer science, to name just a few examples. Traditionally though, data have been scarce and their compilation was subject to controlled collection and deliberate analytical processes (Kitchin 2014a; boyd 2010). In contrast, the '[…] challenge of analysing big data is coping with abundance, exhaustivity and variety, timeliness and dynamism, messiness and uncertainty, high relationality, and the fact that much of what is generated has no specific question in mind or is a by-product of another activity.' (Kitchin 2014a, 2)

Already in 2015, The Gartner Group ceased issuing a big data hype cycle and dropped 'big data' from the Emerging technologies hype cycle. A Gartner analyst justified this decision, not on the grounds of the term's irrelevance, but because of big data's ubiquitous pervasion of diverse domains: it '[…] has become prevalent in our lives across many hype cycles.' (Burton 2015) One might say that the '[b]ig data *hype* [emphasis added] is officially dead', but only because '[…] big data is now the new normal' (Douglas 2016). While one may argue that the concept has lost its 'news value' and some of its traction (e.g. for attracting funding and attention more generally), it is still widely used, not least in the field relevant to his book. For these reasons, I likewise still use the term 'big data' when examining developments and cases in public health surveillance. Despite the fact that the hype around big data seems to have passed its peak, much confusion remains about what this term actually means.

In the wake of the big data hype, the interdisciplinary field of *data science* (Mattmann 2013; Cleveland 2001) received particular attention. Already in the 1960s, Peter Naur – himself a computer scientist – suggested the terms 'data science' and 'datalogy' as preferable alternatives to 'computer science' (Naur 1966; see also Sveinsdottir and Frøkjær 1988). While the term 'datology' has not been taken up in international (research) contexts, 'data science' has shown that it has more appeal: As early as 2012, Davenport and Patil even went as far as to call data scientist 'the Sexiest Job of the 21st Century'. Their proposition is indicative of a wider scholarly and societal fascination with new forms of data, ways of retrieval and analytics, thanks to ubiquitous digital technology.

More recently, data science has often been defined in close relation to corporate uses of (big) data. Authors such as Provost and Fawcett state, for instance, that defining '[…] the boundaries of data science precisely is not of the utmost importance' (2013, 51). According to the authors, while this may be of interest in an academic setting, it is more relevant to identify common principles '[…] in order for data science to serve business effectively' (51). In such contexts, big data are indeed predominantly seen as valuable commercial resources, and data science as key to their effective utilisation. The possibilities, hopes, and bold promises put forward for big data have also fostered the interest of political actors, encouraging policymakers such as Neelie Kroes, European Commissioner for the Digital Agenda from 2010 until 2014, to reiterate in one of her speeches on open data: 'That's why I say that data is the new oil for the digital age.' (Kroes 2012)

There are various ways and various reasons to collect big data in corporate contexts: social networking sites such as Facebook document users' digital interactions (Geerlitz and Helmond 2013). Many instant messaging applications and email providers scan users' messages for advertising purposes or security-related keywords (Gibbs 2014; Wilhelm 2014; Godin 2013). Every query entered into the search engine Google is documented (Ippolita 2013; Richterich 2014a). And not only users' digital interactions and communication, but their physical movements and features are turned into digital data. Wearable technology tracks, archives and analyses its owners' steps and heart rate (Lupton 2014a). Enabled by delayed legal interference, companies such as 23andMe sold personal genomic kits which customers returned with saliva samples, i.e. personal, genetic data. By triggering users' interest in health information based on genetic analyses, between 2007 and 2013, the company built a corporately owned genotype database of more than 1,000,000 individuals (see Drabiak 2016; Harris, Kelly, and Wyatt 2013a; 2013b; Annas and Sherman 2014).[3]

One feature common to all of these examples is the emergence of large-scale, continuously expanding databases. Such databases allow for insights into, for example, users' (present or future) physical condition; the frequency and (linguistic) qualities of their social contacts; their search preferences and patterns; and their geographic mobility. Broadly speaking, corporate big data practices are aimed at selling or employing these data in order to provide customised user experiences, and above all to generate profit.[4]

Big data differ from traditional large-scale datasets with regards to their volume, velocity, and variety (Kitchin 2014a, 2014b; boyd and Crawford 2012; Marz and Warren 2012; Zikopoulos et al. 2012). These 'three Vs' are a commonly quoted reference point for big data. Such datasets are comparatively flexible, easily scalable, and have a strong indexical quality, i.e. are used for drawing conclusions about users' (inter-)actions. While volume, velocity, and variety are often used to define big data, critical data scholars such as Deborah Lupton have highlighted that '[t]hese characterisations principally come from the worlds of data science and data analytics. From the perspective of critical data researchers, there are different ways in which big data can be described and conceptualised' (2015, 1). Nevertheless, brief summaries of the 'three Vs' will be provided, since this allows me to place them in relation to the perspectives of critical data studies.

Volume, the immense scope of digital datasets, may appear to be the most evident criterion. Yet, it is often not clear what actual quantities of historic, contemporary, and future big data are implied.[5] For example, in 2014, the corporate service provider and consultancy International Data Corporation predicted that until 2020 'the digital universe will grow by a factor of 10 – from 4.4 trillion gigabytes to 44 trillion. It more than doubles every two years' (EMC, 2014). How these estimations are generated is, however, often not disclosed. When the work on this chapter was started in January 2016, websites such as *internet live stats* claimed that 'Google now processes over 40,000 search queries every second on average (visualize them here), which translates to over 3.5 billion searches per day and 1.2 trillion searches per year worldwide' (Google Search Statistics, 2016). In order to calculate this estimation, the site draws on several sources, such as official Google statements, Gigaom publications and independent search engine consultancies, which are then fed into a proprietary algorithm (licensed by *Worldometers*). Externally, one cannot assess for certain how these numbers have been calculated in detail, and to what extent the provided information, estimations and predictions may be reliable. Nevertheless, the sheer quantity of this new form of data contributes to substantiating related claims regarding its relevance and authority.

As boyd and Crawford argue, the big data phenomenon rests upon the long-standing myth '[…] that large data sets offer a higher form of intelligence and knowledge that can generate insights that were previously impossible, with the aura of truth, objectivity, and accuracy' (2012, 663). This has fostered the emergence of a 'digital positivism' (Mosco 2015) promoting the epistemological assumption that we can technologically control big data's collection and analysis, to the extent that these data may 'speak for themselves' and become inherently meaningful.

This is especially relevant, since these large quantities of data and their interpretation are closely related to promises about profits, efficiency and bright future prospects.[6] Big data – as wider phenomena, and with regards to respective cases – are staged in certain ways. The possibilities and promises associated with the term are used to signify its relevance for businesses (see e.g. Marr 2015; Pries and Dunnigan 2015; Simon 2013; Ohlhorst 2012) and governmental institutions (Kim, Trimi, and Chung 2014; Bertot et al. 2014), and their need to take urgent action. However, despite such claims for its relevance, the collection and analysis of big data is often opaque. This performative aspect of big data, combined with the common blackboxing of data collection, quantitative methods and analysis, is also related to the frequently raised accusation that the term is to a large extent hyped (Gandomi and Haider 2015; Uprichard 2013; Fox and Do 2013).

Apart from the recurring issue that most big data practices take place behind closed curtains and that results are difficult to verify (Driscoll and Walker 2014; Lazer et al. 2014), the problem of assessing actual quantities is also closely related to big data's *velocity*. Their continuous, often real-time production creates an ongoing stream of additional input. Not only does the amount of data produced by existing sources grow continuously, but as new technologies enter the field, new types of data are also created. Moreover, changes in users' behaviour may alter data not only in terms of their quantity, but also their quality and meaningfulness.

Regarding the *variety* or qualitative aspects of big data, they consist in a combination of structured, unstructured and semi-structured data. While structured data (such as demographic information or usage frequencies) can be easily standardised and, for example, numerically or alphabetically defined according to a respective data model, unstructured and semi-structured data are more difficult to classify. Unstructured data refer to visual material such as photos or videos, as well as to text documents which are/were too complex to systematically translate into structured data. Semi-structured data refer to those types of material which combine visual or textual material with metadata that serve as annotated, structured classifiers of the unstructured content.

The possibilities and promises associated with big data have been greeted with notable enthusiasm: as indicated before, this does not only apply to corporations and their financial interests, but has also been noticeable in scientific research (Tonidandel, King, and Cortina 2016; Mayer-Schönberger and Cukier 2013; Hay et al. 2013). This enthusiasm is often grounded in the assumption that data can be useful and beneficial, if we only learn how to collect, store and analyse them appropriately (Finlay 2014; Franks 2012). Related literature mainly addresses big data as practical, methodological and technological challenge, seeing them as assets to research, rather than as a societal challenge. The main concern and aim of this literature is an effective analysis of such data (see e.g. Assunção et al. 2015; Jagadish et al. 2014). Such positions have, however, been called into question and critically extended by authors engaged in critical data studies.

Critical Data Studies

Current corporate or governmental big data practices, and academic research involving such data, are predominantly guided by deliberations regarding their practicability, efficiency and optimisation. In contrast, approaches in critical data studies are not primarily concerned with practical issues of data usability, but scrutinise the conditions for contemporary big data collection, analysis and utilisation. They challenge big data's asserted 'digital positivism' (Mosco 2015), i.e. the assumption that data may 'speak for themselves'.

Critical data studies form an emerging, interdisciplinary field of scholars reflecting on how corporations, institutions and individuals collect and use 'big' data – and what alternatives to existing approaches could look like. Currently, critical data studies predominantly evaluates social practices involving (big) data, rather than operationalising approaches for research using big data. It mainly encompasses research *on* big data, focused on assessments of historical or ongoing big data projects and practices (Mittelstadt and Floridi 2015; Lupton 2013; boyd and Crawford 2012). Such an approach is also taken in this book.

In addition, some researchers have critically engaged and experimented with research *with* big data. For example, this has been done by using data processing software like *Gephi* in order to show how algorithms and visualisation may influence research results. Importantly, research groups such as the *Digital Methods Initiative* explore the possibilities and boundaries of applying and developing quantitative digital tools and methodologies.[7] However, at present, *critical data studies* predominantly refers to the critique of recent big data approaches. As Mosco points out: 'The technical criticisms directed at big data's singular reliance on quantification and correlation, and its neglect of theory, history, and context, can help to improve the approach, and perhaps research in general – certainly more than the all-too-common attempts to fetishize big data.' (Mosco 2015, 205–206) Therefore, in order to rethink how big data are being used (especially in research), it is also desirable that future approaches are informed by critical data studies perspectives, rather than being analysed subsequently.[8]

Also, without using the umbrella term 'critical data studies', various authors have of course nevertheless critically evaluated the collection and analysis of digital user data. These perspectives emerged in parallel with technological developments that allowed for new forms of data collection and analysis. Critical positions also surfaced with regards to the use of big data in research. In 2007, the authors of a *Nature* editorial emphasised the importance of trust in research on electronic interactions, and voiced concern about the lack of legal regulations and ethical guidelines:

> 'For a certain sort of social scientist, the traffic patterns of millions of e-mails look like manna from heaven. […] Any data on human subjects inevitably raise privacy issues (see page 644), and the real risks of abuse of such data are difficult to quantify. […] Rules are needed to ensure data can be safely and routinely shared among scientists, thus avoiding a Wild West where researchers compete for key data sets no matter what the terms.' (Nature Editorial 2007)

This excerpt refers to familiar scientific tensions and issues that were early on flagged with regards to big data research.[9] Scholars are confronted with methodological possibilities whose risks and ethical appropriateness are not yet clear.

This uncertainty may, however, be 'overpowered' by the fact that these data allow for new research methods and insights, and are advantageous for researchers willing to take the risk. While certain data may be technically accessible, it remains questionable if and how researchers can ensure, for instance, that individuals' privacy is not violated when analysing new forms of digital data. *If* scientists can gain access to certain big data, this does not ensure that using them will be ethically unproblematic. More importantly, the 'if' in this sentence hints at a major constraint of big data research: a majority of such data can only be accessed by technology corporations and their commercial, academic or governmental partners. This issue has been by Andrejevic (2014) the 'big data divide', and has also been addressed by boyd and Crawford, who introduced the categories of 'data rich' and 'data poor' actors (2014, 672ff.; see also Manovich 2011, 5).

Today, globally operating internet and tech companies decide which societal actors may have access to data generated via their respective platforms, and define in what ways they are made available. Therefore, in many cases, scholars cannot even be sure that they have sufficient knowledge about the data collection methods to assess their ethical (in-)appropriateness. This does not merely mean that independent academics cannot use these data for their own research, but it also poses the problem that even selected individuals or institutions may not be able to track, assess and/or communicate publicly how these data have been produced.

The need for critical data studies was initially articulated by critical geography researchers (Dalton and Thatcher 2014; Kitchin and Lauriault 2014) and in digital sociology, with particular regards to public health (Lupton 2014c, 2013). In geographic research this urge was influenced by developments related to the 'geospatial web'. In 2014, Kitchin and Lauriault reinforced the emergence and discussion of critical data studies, drawing on a blog post published by Dalton and Thatcher earlier that year. The authors depict this emerging field as 'research and thinking that applies critical social theory to data to explore the ways in which they are never simply neutral, objective, independent, raw representations of the world, but are situated, contingent, relational, contextual, and do active work in the world' (Kitchin and Lauriault 2014, 5). This perspective corresponds to Mosco's critique that big data 'promotes a very specific way of knowing'; it encourages a 'digital positivism or the specific belief that the data, suitably circumscribed by quantity, correlation, and algorithm, will, in fact, speak to us' (Mosco 2015, 206). It is exactly this digital positivism which is challenged and countered by contributions in critical data studies.

When looking at the roots of critical data studies in different disciplines, one is likely to start wondering which factors may have facilitated the development of this research field. In the aforementioned blog post 'What does a critical data studies look like, and why do we care?' Dalton and Thatcher stress the relevance of geography for current digital media and big data research, by emphasising that most information nowadays is geographically/spatially annotated (with reference to Hahmann and Burghardt 2013). According to the authors, many of the tools and methods used for dealing with and visualising large amounts of digital data are provided by geographers: 'Geographers are intimately involved with this recent rise of data. Most digital information now contains some spatial component and geographers are contributing tools (Haklay and Weber 2008), maps (Zook and Poorthius 2014), and methods (Tsou et al. 2014) to the rising tide of quantification.' (Dalton and Thatcher 2014)

Kitchin and Lauriault explore how critical data studies may be put into practice. They suggest that one way to pursue research in this field is to '[…] unpack the complex assemblages that produce, circulate, share/sell and utilise data in diverse ways; to chart the diverse work they do and their consequences for how the world is known, governed and lived-in' (Kitchin and Lauriault 2014, 6). Already in *The Data Revolution* (2014a), Kitchin suggested the concept of data assemblages. In this publication, he emphasises that big data are not the only crucial development in the contemporary data landscape: at the same time, initiatives such as the digital processing of more traditional datasets, data networks, and the open data movement contribute to changes in how we store, analyse, and perceive data. Taken together, various emerging initiatives, movements, infrastructures, and institutional structures constitute data assemblages that shape how data are perceived, produced and used (Kitchin 2014a, 1)

By drawing on the same idea of digital data assemblages, Lupton outlines a critical sociology of big data (2014b, 93). The author conceptualises big data as knowledge systems which are embedded in and constitute power relations. In a first step, she examines the various fields of their utilisation, such as humanitarian uses, education, policing and security. Moreover, she deconstructs the metaphors which were initially used to describe big data, and how these reflect contemporary criticism. Terms such as 'trails', 'breadcrumbs', 'exhaust', 'smoke signals', and 'shadows' (Lupton 2014b, 108) indicate that big data are commonly seen as signs with a strong indexical quality. The latter part of her analysis also provides an initial overview of themes in the field of critical data studies. However, only in a later online publication (Lupton 2015) does Lupton use the term 'critical data studies'.

A crucial metaphor that Lupton refers to here is the notion of 'raw data' (Boellstorff and Maurer 2015; Gitelman 2013; Boellstorff 2013). The rejection of an idea of data as implicitly 'natural' and 'given', i.e. 'raw', is a crucial tenet in critical data studies. Drawing on Lévi-Strauss's 'culinary triangle' of *raw-cooked-rotten* as well as Geertz' methodological approach and genre of *thick descriptions*, Boellstorff (2013) criticises the nature-culture opposition which is implied in the differentiation between 'raw' (collected) and 'cooked' (processed) data. Rather than being 'pure' expressions of human behaviour or opinions, data in all their manifestations, are always subject to interpretation and normative influences of meaning-making. To frame this fundamental condition of data-driven processes, the author suggests the notion of 'thick data': 'what makes data 'thick' is recognizing its irreducible contextuality: 'what we inscribe (or try to) is not *raw* social discourse.' […] For Geertz, 'raw' data was already oxymoronic in the early 1970s: whether cooked or rotted, data emerges from regimes of interpretation' (Boellstorff 2013).

The idea of rotten data pursues the metaphor of 'raw' and 'cooked' data, but calls attention to the changes in data and their accessibility which go beyond technically or methodologically intended control. Boellstorff (2013) argues that 'the 'rotted' 'allows for transformations outside typical constructions of the human agent as cook—the unplanned, unexpected, and accidental. Bit rot, for instance, emerges from the assemblage of storage and processing technologies as they move through time.'

In a later publication, Boellstorff and Maurer (2015) identified 'relation' and 'recognition' as particularly crucial factors influencing the constant process of data interpretation – which starts with its selection and collection. Data are created and given meaning in interactions between human and non-human actors. Their recognition is socio-culturally and politically defined (Boellstorff and Maurer 2015, 1-6; see also Lupton 2015). In this sense, the term data, derived from the Latin plural of datum, 'that is given', is already misleading, and indicates the term's socially constructed meaning. Strictly speaking, '[o]ne should never speak of 'data'- what is given – but rather of sublata, that is, of 'achievements." (Latour 1999, 42)

It is not surprising that many of the critical approaches to big data are related to fields in which potentially derived information is inherently rather sensitive: in health research and with regards to location-based technology, data critique has emerged as an important general theme. So, the need for critical data studies goes beyond such fields, and should engage with data which have been traditionally seen as sensitive, i.e. allowing for access to information which is commonly treated as private or confidential. One challenge for critical data studies has been (and will be) to demonstrate to what extent seemingly impersonal data are in fact highly sensitive, due to, for example, their corporate, regulatory or technological embedding, and new means for interrelating datasets.

Aims and Chapters

More generally, the aim of this book is to contribute to the emerging field of critical data studies. Specifically, it does so by examining the implications of big data-driven research for ethico-methodological decisions and debates. I analyse how research in public health surveillance that involves big data has been presented, discussed and institutionally embedded. In order to do so, I will examine projects employing and promoting big data for public health research and surveillance.[10] This book realises three main objectives: first, it develops and applies a critical data studies approach which is predominantly grounded in pragmatist ethics as well as Habermasian discourse ethics, and takes cues from (feminist) technoscience criticism (Chapter 2). Second, it identifies broader issues and debates concerning big data-driven biomedical research (Chapter 3). Thirdly, it uses the example of big data-driven studies in public health research and surveillance to examine more specifically the issues and values implicated in the use of big data(Chapters 4 and 5).

This book is divided into six chapters. Chapter 1 introduced the term 'big data' and provided an initial overview of critical data studies. Chapter 2 'Examining data practices and data ethics' focuses on the theoretical foundations of my analysis. The first subchapter 'What it means to "study data" expands on the brief introduction to critical data studies provided above. Adding to the basic principles and historical development outlined in Chapter 1, it offers an overview of themes and issues. The second subchapter, 'Critical perspectives' elucidates why the approach taken in this book should be considered 'critical'. While Habermas' work links this book to critical theory, I also draw on strands in science and technology studies which have explicitly addressed the possibilities and need for normative, engaged analyses; here, I refer mainly to the sociology of scientific knowledge construction, as well as feminist technoscience. The third subchapter on pragmatism and discourse ethics builds upon

Keulartz et al.'s pragmatist approach and Habermas' critical theory notion of discourse ethics.

Chapter 3 'Big data: Ethics and values' describes normative developments which have been discussed with regards to digital data practices, particularly in research. This chapter depicts tensions between values related to personal rights and those linked to the public good, such as the common opposition between privacy and security. Moreover, it shows how transparency and open data relate to (and may conflict with) individuals' privacy and corporate interests in exclusive data access. Based on an overview of the values which have been advanced to justify or critique big data research, I examine how these relate to current negotiations of research methodologies and normativities. This also involves reflections on entanglements between corporate data economies and research analytics. The main purpose of this chapter is to identify broader developments relevant to the case studies, as well as those values which have been comparatively emphasised or neglected.

Chapters 4 and 5 examine the institutional context, methodological choices and justifications of big data-driven research in public health surveillance. In Chapter 4, I show how funding schemes specifically targeted at promoting the use of big data in biomedical research incentivise methodological trends, with ethical implications. Interdependencies between researchers and grant providers need to be seen in the context of funding environments which are partly co-defined by internet/tech corporations. Shedding light on these institutional contexts also facilitates insights into factors co-constructing researchers' decisions to pursue certain topics and approaches.

Chapter 5 goes on to show how such research decisions and developments are translated into research projects. I specifically unpack how the use of big data collected by tech corporations is practically realised as well as discursively presented by researchers. I focus on research projects which have utilised sources that are not traditionally seen as 'biomedical data', but should be seen as such since they allow for insights into users' state of health and health-relevant behaviour. Analyses of specific cases and references to contemporary developments are made throughout the book, but especially in Chapter 5.

While Chapter 4 highlights the institutional conditions for public health surveillance involving digital data by depicting relevant funding schemes, Chapter 5 presents three clusters of cases: 1) Tweeting about illness and risk behaviour; 2) data retrieval through advertising relations; and 3) data mashups. The first cluster examines how Twitter data have been utilised as indicators of health risk behaviour. The second cluster explores researchers' attempts to access, for example, Facebook data via advertising and marketing services. The third cluster focuses on publicly available platforms developed by researchers which draw on data collected by tech corporations such as Google.

These case studies have been chosen because they are not merely clear-cut cases of corporate, commercial data utilisation, but involve more diverse values. More importantly, they are cases in which the analytic possibilities of big data have led to the emergence of 'technosciences', i.e. academic research fields which are substantially grounded in technological changes. It seems important to highlight here that the book's objective is not merely to expose certain projects as 'immoral' (see also Chapter 2). Instead, I want to emphasise the complexities and contradictions, the methodological as well institutional dilemmas, and factors of influence co-constructing current modes of big data research.

The final chapter ties together insights from the analysis, specifically in relation to the critical perspectives and theory introduced in Chapter 2. It emphasises two main issues: first, in the field of big data-driven public health research, one can observe complex (inter-)dependencies between academic research and the commercial interests of internet and tech corporations. This is notably related to two main developments: on the one hand, data access is often controlled by these companies; on the other hand, these companies incentivise research at the intersection of technology and health (e.g. through funding and selective data access).

Second, data practices, norms and the promises of internet/tech corporations are increasingly echoed and endorsed in big data-driven health research and its ethics. These tendencies foster research approaches that inhibit the discursive involvement of affected actors in negotiations of relevant norms. In consequence, I argue that, from a discourse ethics perspective, there is an urgent need to transition from big data-driven to data-discursive research, foregrounding ethical issues. Such research needs to encourage the involvement of potentially affected individuals, as a condition for formative discourse and research ethics grounded in valid social norms.

CHAPTER 2

Examining (Big) Data Practices and Ethics

When adding the term 'critical' to an analysis or field of research, one may be inclined to intuitively associate this primarily with critical theory. And of course, critical theory is a decisive field of research for investigations concerned with matters of power and societal inequalities (see also Fuchs 2014, 7ff.; Feenberg 2002). However, this is not the only research line which is crucial for an understanding of *critical* research.

This chapter elaborates on the critical, theoretical foundations and approach of this book. First, following up on the initial overview of critical data studies (CDS), I take a closer look at what it means to 'study data'. Second, given that this book is part of the series *Critical Digital and Social Media Studies* and draws on CDS, I will reflect on what it means to pursue a critical stance and approach. The subchapter 'Critical perspectives' pays attention to links between critical data studies and concepts rooted in poststructuralism, as well as the philosophy of science.

Since I investigate research involving big data, their conditions and ethical implications, my analysis likewise draws on insights and debates in science and technology studies (STS). Due to the critical, i.e. normative perspective and attention to power relations, I am particularly interested in the relevance of political issues in STS, as well as the possibilities and constraints for making normative arguments in this field. While STS has often been criticised for its lack of political engagement and merely disguised normativity, I discuss how certain branches and debates have embraced critical, normatively engaged perspectives. This argument will be underlined in relation to the 1990s debate on the politics of SSK, i.e. the sociology of scientific knowledge production (Radder 1998; Richards and Ashmore 1996; Wynne 1996). I also take some cues from feminist technoscience (Wajcman 2007; Weber 2006; Haraway 1997).

This broader contextualisation leads up to the main theoretical foundation of my approach. I draw on Keulartz et al.'s (2004) pragmatist approach

---

How to cite this book chapter:
Richterich, A. 2018. *The Big Data Agenda: Data Ethics and Critical Data Studies.*
Pp. 15–31. London: University of Westminster Press.
DOI: https://doi.org/10.16997/book14.b. License: CC-BY-NC-ND 4.0

to ethics for my analysis of big data-driven research practices. Conceptually, particular emphasis is put on Habermas' theory of 'discourse ethics' (2001 [1993]; 1990; see also Rehg 1994.). In employing this concept, my analysis is likewise informed by Habermas' contribution to critical theory. While rejecting techno-deterministic as well as substantive views[11], I unpack interdependencies between technological developments, corporate data practices and big data-driven health research, specifically in the field of public health surveillance. In consequence, this book inevitably grapples with emerging power asymmetries (Sharon 2016; Andrejevic 2014) and questions of data (in-)justice (Taylor 2017; Dencik, Hintz and Cable 2016; Heeks and Renken 2016) crucial to CDS.

The critical perspectives, theories and approach outlined in this chapter make a much-needed contribution to the field of big data-driven health research. They allow us to view ongoing big data practices and discourses in a different light, nuancing and challenging influential, taken for granted claims grounded in digital positivism. Such contributions are necessary to facilitate debates and decision-making processes which consider the advantages, disadvantages and alternatives, the realistic possibilities, risks and uncertainties of big data.

What it Means to 'Study Data'

The term critical data studies (CDS), very plainly, suggests two things: first, that scholars working in this field investigate data; second, that they do so from critical perspectives. When focusing initially on the latter part of this umbrella term, one may ask what it means to 'study data'. What kinds of subjects and approaches are examined in this field? Studying data in this context does not merely imply utilising or analysing 'data as such'.[12] Instead, CDS interrogates the embeddedness of data in (knowledge) practices, institutions, and political and economic systems. In some cases, this might be done by reflectively experimenting with big data utilisation, but critical data research goes beyond mere quantitative analyses of data. Instead, it qualitatively questions their constructedness, affordances and implications. CDS scholars examine the complex interplay between data and the institutions and actors that produce, own and utilise them. They might for example discuss: how social networks such as Facebook draw on user data (Oboler, Welsh, and Cruz 2012); how big data are utilised in the food and agriculture sectors (Bronson and Knezevic 2016); how genomic data arise from digital (corporate) services (Harris, Kelly, and Wyatt 2016); or how data brokers retrieve and monetise individuals' data (Crawford 2014).

The relevance of *justice* in relation to data has been – implicitly and explicitly – a key concern for critical data studies. For example, drawing on prior work on 'information justice' (Johnson 2014) and 'data justice' (Heeks and Renken 2016), Taylor suggests a framework centred on ensuring just data practices. It is aimed at countering marginalisation as well as power asymmetries, and at facilitating just approaches to data retrieval and use. In consequence, her

'[…] approach begins not from a consideration of the average person, but asks instead what kind of organising principles for justice can address the marginalised and vulnerable to the same extent as everyone else' (Taylor 2017, 20).

This focus on justice is also implicitly expressed in critical data studies' broader concern with power relations and the agency of key stakeholders. Analyses may focus, for instance, on governmental big data practices (see e.g. Rieder and Simon 2016; Lyon 2014; Van Dijck 2014; Tene and Polonetsky 2012), corporate data retrieval, analysis, and use (see e.g. Bronson and Knezevic 2016; Lazer et al. 2014; Oboler, Welsh, and Cruz 2012) or big data-driven research in universities and non-profit institutions (see e.g. Borgman 2015; Gold and Klein 2016; Kaplan 2015; Franke et al. 2016, Wyatt et al. 2013, Kitchin 2013). For example, due to the dominance of media corporations in retrieving user-generated big data, research institutions are increasingly dependent on access conditions defined by these companies. And while governments are trying to regulate corporate data collection (European Commission 2014), we have likewise witnessed severe violations of users' privacy and attempts to integrate corporate data in governmental surveillance (see e.g. Lyon 2014; Van Dijck 2014). Overlaps, collaborations, competition and conflicts emerge between actors in these different, entangled areas. Similarly, by focusing on big data use in public health surveillance, this book calls attention to interdependencies between corporate big data practices, scientific research and its ethics.

As Lupton points out: 'While critical data studies often focuses on big data, there is also need for critical approaches to 'small' or personal data, the type of information that people collect on themselves.' (2014, 4). This criticism has now been partly addressed, thanks to Lupton's own work as well as more recent contributions to CDS (see e.g. Sharon and Zandbergen 2016; Milan 2016; Schrock 2016). This requirement is likewise considered in this book, even though I argue that small, personal data are often inseparable from big, corporate data. A main reason for this is that individuals' potential to collect data individually is commonly tied to sharing commitments which are difficult or impossible to avoid. On the one hand, we should not forget that corporate, governmental, and scientific big data practices predominantly rely on information generated by individuals. On the other hand, these users should indeed not merely be 'victimised' – despite the importance of power asymmetries in big data utilisation. Instead, one also needs to acknowledge those practices through which individuals engage critically and actively with data.

As mentioned at the beginning of this subchapter, CDS stresses the embeddedness of big and small data, and the need for context sensitivity. In this sense, research in this field resembles sub-disciplines of digital media and internet studies, such as *software studies* (Manovich 2013; Berry 2011a; Kitchin and Dodge, 2011; Fuller 2003), *critical algorithm studies* (Kitchin 2017; Gillespie and Seaver 2015), and *platform studies* (Bogost and Montford 2009). Software, platform, and algorithm studies all emphasise the need to analyse computational objects and practices, not merely as technical, but as social issues. They highlight the necessity to look beyond matters of content and to investigate the interplay between technological intricacies and social, political, and economic factors. This aim is often explicitly related to scholars such as Friedrich Kittler, Michel Foucault, Gilles Deleuze and Félix Guattari.

Kittler's work is commonly cited, since he early on theorised the interplay between software and hardware, emphasising the need for a 'proper understanding of the science and engineering realities that govern the highly fine-structured computer worlds in which we live' (Parikka 2015, 2). Fuller, among others, draws on Deleuze and Guattari's work in arguing that '[…] software constructs ways of seeing, knowing, and doing in the world that at once contain a model of that part of the world it ostensibly pertains to and that also shape it every time it is us' (Fuller 2003, 19). Similarly, algorithms are described as 'Foucauldian statements' through which 'historical existence accomplishes particular actions' (Goffey 2008, 17). More generally, software and algorithm studies alike are often linked to Foucault's conception of power, not as force which is exerted on individuals or groups, but as a dynamic embedded in and permeating societies (Foucault 1975). Similarly, such theoretical foundations tend to be crucial for the critical perspectives developed in CDS.

Critical Perspectives

Critical data studies is a field that acknowledges and reflects on the practices, cultures, politics and economies unfolding around data (Dalton, Taylor, and Thatcher 2016). Issues addressed in this field may range from the abovementioned themes such as individuals' privacy and autonomy, to data science ethics and institutional changes triggered by corporate or governmental funding invested in big data research. All these perspectives have in common that they highlight the need for analyses of big data practices which are conscious of power relations, biases, and inequalities. Likewise, they are open to an empirical engagement with societies permeated by digital data.

When reflecting on what it means – or should mean – to conduct *critical* data studies, Dalton, Taylor, and Thatcher advise caution in defining this attribute. They point out that a narrow definition restricting critical research to the domain of normative, critical theory would be counterproductive: 'When you append 'critical' to a field of study, you run the risk of both offending other researchers, who rightly point out that all research is broadly critical and of bifurcating those who use critical theory from those who engage in rigorous empirical research' (Dalton, Linnett, and Thatcher 2016).

So far, in CDS, the notion of 'criticalness' has frequently been grounded in poststructuralist theory, and in some cases established with reference to the philosophy of science. In their chapter 'Data Bite Men' (2014), Ribes and Gitelman coin the term 'commodity fictions of data' (147). Referring to Foucault's 'commodity fiction of power' (165), they aim to '[…] reveal the complex assemblage of people, places, documents, and technologies that must be held in place to produce scientific data' (147). While Kitchin and Lauriault (2014) base their critical approach inter alia on Foucault's notion of *assemblages* and *dispositive*, they also draw on the work of science philosopher Ian Hacking. Likewise, Symons and Alvarado stress the relevance of philosophy of science for CDS. The authors argue that '[t]he assumptions governing the atheoretical turn are false and, as we shall see, studying Big Data without taking contemporary philosophy of science into account is unwise [...]' (2016, 2). Despite its potential for valuable contributions to CDS, the authors describe philosophy of science as a disregarded approach to the field so far (ibid.)

It is of course far from surprising, and perfectly valid, that a variety of academic perspectives claims to 'be critical'. In this sense, it is also not a clearly defined set of theories which is defining for the aims and possibilities of CDS. Instead, scholars in this field explore and develop multiple theories embedded in datafication. In doing so, they respond to the shared concern that unreflectively embracing technological changes related to (big) data may hinder sustainable and just techno-social developments. They do not *assume* that changes associated with big data are risky or harmful, but they scrutinise the possibility that they could be. Among the common tenets of CDS are the following assumptions, which likewise define how this book qualifies as 'critical':

- *Data politics and agency:* Data are not neutral. They have agency and they express the agency (or lack thereof) of related actors (Iliadis and Russo 2016; Crawford, Gray and Miltner 2014).
- *Data economies and ownership:* Data may be produced by many, but they are controlled by a few, often corporate, actors (Andrejevic 2014).
- *Data epistemologies:* Big data are as constructed as any other form of information and knowledge, but claims regarding their inherent superiority have contributed to a 'digital positivism' (Mosco 2015; see also Symons and Alvarado 2016).

Essentially, these assumptions highlight interdependencies between emerging technologies and (human) actors in increasingly datafied societies. Big data are as much a product of contemporary socio-technical conditions, as they are producers of such conditions. This last point reflects the idea of co-construction, which has long been a crucial concept in *science and technology studies* (STS). In the mid-1980s, the social construction of technology (SCOT) approach (Pinch and Bijker 1984) stressed that users are not simply passive receivers, but play a role in defining the meanings, successes and failures of technologies. In describing 'the mutual shaping of social groups and technologies' (Oudshoorn and Pinch 2003, 3), the notion of co-construction acknowledges that techno-social developments are neither imposed on societies nor are technological changes implemented by human actors in an entirely controlled manner (Bijker 1995).

The tenet of co-construction aims at avoiding techno-deterministic as well as substantive views. Yet one should not forget that the initial assumptions of SCOT have been criticised and revised (including by the authors themselves, see e.g. Pinch 1996; Bijker 1995): 'A central target of criticism is SCOT's view of society as composed of groups. [...] Implicitly, SCOT assumes that groups are equal and that all relevant social groups are present in the design process. This fails to adequately attend to power asymmetry between groups.' (Klein and Kleinman 2002, 30). Thus, co-construction needs to factor in the barriers to and inequalities in decision making, implementation, and acceptance of emerging technologies.

It needs to be considered that STS has a rather ambiguous relation to normative assessments of technology. At least historically, STS has been dominated by an emphasis on 'neutrality' and 'descriptiveness' (Radder 1998; Richards and Ashmore 1996). This lack of (open) normativity has also been criticised as an obstacle when it comes to political implications and necessary decisions, typical for the context of technological developments and establishments (Law 2008). Despite this tendency in earlier strands of STS, an understanding of co-construction which accounts for power imbalances can be highly valuable for a critical analysis of big data practices. It allows for a nuanced understanding of the role of actors involved in and affected by big data utilisations. Oudshoorn and Pinch (2003) emphasise the importance of neither over- nor underestimating actors' agency in technological cultures:

> [T]he co-construction of users and technologies may involve tensions, conflicts, and disparities in power and resources among the different actors involved. [...] we aim to avoid the pitfall of what David Morley (1992) has called the 'don't worry, be happy' approach. A neglect of differences among and between producers and users may result in a romantic voluntarism that celebrates the creative agency of users, leaving no room for any form of critical understanding of the social and cultural constraints on user-technology relations. (16)

Acknowledging this aspect of co-construction is likewise relevant to CDS. A critical analysis of data practices requires an assessment of the interplay between human practices, institutional constellations, technological developments, and the agencies embedded and implicated within these actors.

Within STS, certain strands are particularly concerned with the conundrum of descriptive versus explicitly normative approaches to technology assessment. The quote above from Oudshoorn and Pinch is a first indication of more critical perspectives dealing with constraints, biases, and (power) imbalances. Historically, it seems especially relevant to highlight the late 1990s debate on the politics of SSK: the sociology of scientific knowledge production (Radder 1998; Richards and Ashmore 1996; Wynne 1996). The negotiations resulting from a special issue on this topic may be seen as a milestone for voicing normativity and politics in STS. Key insights of this debate are relevant to this book, since such a critical, i.e. politically and economically conscious, perspective regarding the production of scientific knowledge is likewise needed to assess the knowledge claims posed by big data research.

Starting with the telling sub-heading 'If You Can't Stand the Heat...', Richards and Ashmore argue in their article that '[t]he question of whether the sociology of scientific knowledge (SSK) can be, should be, or inescapably is, 'political' is one that has been with us since its inception in the early 1970s' (Richards and Ashmore 1996, 219). As editors of a special issue of *Social Studies of Science* on 'The Politics of SSK', they brought together papers which negotiate 'commitment versus neutrality in the analysis of contemporary scientific and technical controversies' (220). The included articles deal with the political implications of scientific knowledge production and assessment.

While the special issue also includes defences of the need for 'neutral social analysis' (Collins 1995), it provides notably rich insights into the risks of neglecting political issues in scientific knowledge production. In contrast to the (back then) common STS 'ideal of a 'value-free' relativism' (Pels 1996, 277), Pels calls for the acknowledgement of 'third positions' in assessments of scientific knowledge production which '[...] are not external to the field of controversy studied, but are included and implicated in it. [...] They are not value-free or dispassionate but situated, partial and committed in a knowledge-political sense.' (282). In this sense, the aim of my analysis is to be critical by being not only 'normatively relevant, but also normatively engaged' (Radder 1998, 330).

Such an approach appears to be a necessary contribution to current debates regarding research on and with big data, since their societal benefits and potential have been widely overemphasised. I see striking parallels between the 'early era of big data' and the historical context during which the abovementioned special issue 'The Politics of SSK' was published. The editors argue that it was launched at a time when SSK was '[...] under renewed attack from die-hard, positivist defenders of science its hitherto epistemologically-privileged view of the world and people' (Richards and Ashmore 1996, 219). Similarly, the big data hype has been accompanied by claims concerning the obsolescence of theories and hypotheses at a time where data may (allegedly) 'speak for themselves'.

The strengths of situated, partial and committed perspectives – conceptualised by Pels (1996) as an inevitability which one should not disguise – were raised with particular emphasis in feminist technosciences (Harding 2004, 1986; Haraway 1988). Feminist scholars have countered the assumption that relevant technology assessments can and should be symmetric and impartial (Wajcman 2007; Weber 2006). Their work serves as an important reminder that feminist critique likewise applies to how big data are being presented.[13]

For instance, Haraway argued that the common presentation of scientific knowledge as beyond doubt and revision, and allowing for generalisable objectivity, tends to create a 'view of infinite vision' which 'is an illusion, a god trick'

(Haraway 1988, 583). One can clearly see how this illusion of infinite vision and objectivity is revived in current debates on big data. In contrast, and this corresponds to the critical perspective presented in this book, '[…] feminist objectivity is about limited location and situated knowledge, not about transcendence and splitting of subject and object' (Haraway 1988, 583). With regards to big data-driven research and the moralities/norms crucial to the approaches taken, particularly from a discourse ethical perspective, this also points to the question of which standpoints are systematically included or excluded.

In conclusion, in this sub-chapter I have argued that the following assumptions are relevant to my critical analysis of big data research practices. Debates on normativity in STS and the politics of SSK have brought about an idea of socio-technical co-construction, aware of power asymmetries between groups and actors. Taking cues from these early debates, I pursue a critical understanding of societal changes that neither assumes the dominance of technology nor the unimpaired impact of human actors. My perspective is critical and normative in the sense that I pay particular attention to power imbalances, issues of justice, and a potential lack of democratic structures in big data-related research, its communication, and debate. This is also closely connected to issues raised in feminist technoscience, reminding us that techno-scientific developments such as big data commonly echo the claims and promises of powerful actors, while neglecting subjugated positions.

While these are more general principles underlying my analysis, the following chapter on 'Pragmatism and discourse ethics' specifies my approach and the questions relevant to my analysis. I have opted for a pragmatist approach to ethics as proposed by Keulartz et al. (2004), with particular emphasis on Habermasian discourse ethics.[14] The latter concept is particularly relevant, as it establishes justice as a normative cornerstone for discursive conditions under which (valid) social norms are formed.

Approach: Pragmatism and Discourse Ethics

In public discourses, proponents of new technologies articulate promises, and evoke hopes and expectations. In response to such discourses or to evolving socio-technological practices, positions expounding risks and uncertainties may also be brought forward. This, obviously simplified, dynamic applies for example to wearable activity/fitness trackers, an important technology for the retrieval of digital user data.

The popularisation of wearable activity/fitness trackers was accompanied by claims that the use of (and data collection with) these devices would improve users' wellbeing, health and life expectancy. It was also proposed that they would significantly decrease healthcare costs (Chang 2016; 'Wearing Wellness' 2016). It was suggested, for example, that '[…] 56 percent of those with these trackers believe that their average life expectancy has increased by a decade thanks to their ability to monitor their vital signs on a daily basis' (Chang 2016)[15]. Some health insurance providers, for instance in the United States, were quick to react, and offered discounts to those customers who would be willing to provide access to their tracker data (Mearian 2015).

But there is also concern about, and resistance to, the technology's influence on contemporary and future societies. In response to the popularisation of these activity/fitness trackers and their social implications, issues regarding fairness, discrimination, privacy, data abuse and safety were raised (Collins 2016; Liz 2015). Not everyone may be able to afford such a tracker in the first place; health impaired users, especially those suffering from a restriction of motion, are excluded from insurance benefits offered for tracking an active lifestyle.

Boyd (2017) concludes an article by calling on users not to ignore the possibility that data collected via activity trackers may be used to their disadvantage: '[T]he devices could provide justification for denying coverage to the inactive or unhealthy, or boosting their insurance rates. Consumers should not assume their insurance companies will use their data only to improve patient care. With millions of dollars on the line, insurers will be sorely tempted.' Such arguments are typical for discourses surrounding emerging technologies and techno-social practices.

Conceptually, for the context of this book, big data should be understood as an umbrella term for a set of emerging technologies. As Kitchin (2014a) and Lupton (2014b) emphasise, in using the notion of data assemblages we need to account for cultural, social and technological contexts, networks, infrastructures, and interdependences that can make sense of big data. The term 'big data' does not only relate to the data as such, but also to the practices, infrastructures, networks, and politics influencing their diverse manifestations. Understanding big data as a set of emerging technology seems conceptually useful, since it encompasses digitally enabled developments in data collection, analysis, and utilisation.

Key insights regarding the dynamics of emerging technologies are applicable to current big data debates and practices. With regards to nanotechnology, Rip describes the dilemma of technological developments: 'For emerging technologies with their indeterminate future, there is the challenge of articulating appropriate values and rules that will carry weight. This happens through the articulation of promises and visions about new technosciences [...].' (Rip 2013, 192) According to Rip, emerging technologies are sites of 'pervasive normativity' characterised by the articulation of promises and fears. He conceptualises such 'pervasive normativity' as an approach 'in the spirit of pragmatist ethics, where normative positions co-evolve' (2013, 205).[16] We can observe such dynamics in relation to big data too, as with the example of data collection enabled by activity trackers. These have provoked communication, arguments and debates justifying, countering and negotiating their corporate, governmental, institutional and academic/scientific utilisation.

Rip's perspective specifically, and pragmatist ethics more generally, stress that establishing new technologies is not just a matter of bringing forward 'objective arguments' regarding their superiority. Instead, they are introduced into societies in which they are discursively associated with/dissociated from certain norms and values. As indicated above, actors with an interest in the popularisation of a certain technology may that way become involved in encouraging its use. Again, the big data hype and the activity tracker example mentioned above are textbook examples of such dynamics: proponents emphasise values and norms which they deem supportive for paving the way for a technology's acceptance and utilisation (and belittle those seen as adverse).

At the same time, these positions will likely be challenged, opposed and contradicted. Pragmatism, among other fields, reminds us that the rise of big data and related research practices is not a mere matter of their technological superiority. Instead, they form a field of normative justification and contestation. Thus, such a pragmatist approach to ethics – in conjunction with the critical literature introduced in Chapters 1 and 2 – has also been chosen in this book.

As briefly introduced in Chapter 1, I draw on Keulartz et al.'s[17] suggestions for a 'pragmatist approach to ethics in technological cultures' (2004, 14). This approach has been developed not as '[…] a complete alternative for other forms of applied ethics but rather a complement', aimed at a '[…] new perspective on the moral and social problems and conflicts that are typical for a technological culture' (Keulartz et al. 2004, 5).[18] The term *technological culture* emphasises the rapid changes and dynamics which individuals experience in postmodern societies. It does not only relate to technological developments as such, but to their influence on and interaction with norms, values and social practices.

(Neo-)pragmatist approaches to ethics accommodate epistemological insights into the fallibility of (scientific) knowledge, while allowing for critical assessments of societal power structures.[19] Keulartz et al. propose their 'pragmatist approach to ethics in a technological culture' (2004) as alternative which combines the strengths of applied ethics and science and technology studies, while avoiding the weaknesses of these fields. According to the authors, applied ethics is an effective approach when it comes to detecting and voicing the normativities implied in or resulting from socio-technical (inter-)actions, but it lacks possibilities to capture the inherent normativity and agency of technologies (Keulartz et al. 2004, 5). While STS implies or allows for these possibilities, most modern STS approaches still suffer from a 'normative deficit' (12) and a rarely contested tendency to insist on descriptive, unbiased analyses.[20]

This concern has already been outlined in the previous sub-chapter, highlighting some of the strands in STS that are committed to critical, normative assessments (see also Winner 1980, 1993). In accordance with such a commitment to critical engagement and normative assessments, Keulartz et al. propose their approach as an attempt to overcome the lack of normative technology assessments. They argue that the 'impasse that has arisen from this', (i.e. the respective 'blind spots' of applied ethics and STS) can 'be broken by a reevaluation of pragmatism' (2004, 14). Pragmatism is rooted in American philosophy, and most notably in the 'classical' works by Charles Sanders Peirce, William James, John Dewey, and George Herbert Mead. Despite the immense diversity of approaches using the label 'pragmatism', according to the Keulartz et al. (2004: 16ff.), these can be characterised by three shared anti-theses and principles: anti-foundationalism, anti-dualism, and anti-scepticism (see also Rorty 1994, 1992).

Anti-foundationalism refers to the principle of fallibilism. Antifoundationalist accounts give up on the possibility that we may reach certainty with regards to knowledge or values, i.e. discover some 'ultimate truth'. Instead, they assume that knowledge, just as much as values and norms, is constantly being renegotiated. This by no means implies, however, that anti-foundationalism rejects the possibility of knowledge or values. Instead, it differentiates between more or less reliable and well-grounded knowledge: in this sense, knowledge is not seen as universal and beyond eventual revision, but may be subject to reconsideration in light of future discoveries or developments. This anti-thesis also implies that moral values are not simply static, but may be renegotiated in relation to technological developments – which may not be simply 'approved', but just as much contested and rejected.

Antidualism stresses the need to refrain from predefined, taken-for-granted dichotomies. Among the criticised dualisms mentioned by Keulartz et al. are essence/appearance, theory/practice, consciousness/reality, and fact/value. Applied ethics tends to assume such dualisms as a priori. In contrast, pragmatism stresses the interrelations and blurred lines between such categories. While it may revert to these categories, it forms them out of empirical material and does not essentially 'apply' them. It assigns merely analytical value to such categories, rather than any ontological status. This anti-thesis also resembles the idea of co-construction, which aims at avoiding a simplistic opposition of technical impact and societal reaction (and vice versa).

Lastly, *anti-scepticism* (and its reconciliation with fallibilism) is a main principle of pragmatism. It is closely linked to the need for situated perspectives and explicit normativity. It refers to the anti-Cartesian foundation of pragmatism: 'We have no alternative to beginning with the 'prejudices' that we possess when we begin doing philosophy. [...] The prejudices are 'things which it does not occur to us can be questioned. [...] Cartesian doubt 'will be a mere self-deception, and not real doubt' (Hookway 2008, 154, citing Peirce). In this sense, we cannot begin with complete doubt, just as we cannot begin with absolute objectivity. Here again, the feminist and SSK insistence on situated knowledge and acknowledgement (as far as possible within these epistemic constraints) of normative values in research practices are crucial.

Pragmatism has been only hesitantly taken up in European research. It was associated with negative 'stereotypes about the land of the dollar' (Joas 1993, 5). It was often dismissed as 'superficial and opportunistic', and accused of 'utilitarianism and meliorism' (Keulartz et al. 2004, 15). In an overview of pragmatism's reception, Joas (1993) contended: 'Disregarding the obviously spectacular exceptions – Karl-Otto Apel and Jürgen Habermas (as well as a few other specialists there) – in Germany, by contrast, pragmatism is even today having a very rough time of it.' (2). But in the late 1990s and 2000s, pragmatism experienced a somewhat unexpected revival and popularisation even in European research (see Keulartz et al. 2004, 15ff.; Baert and Turner 2004; Dickstein 1998). Apart from the influential work of American philosophers such as Hillary Putnam and Richard Rorty, the European popularisation of pragmatism can also be traced back to the impact of the abovementioned Karl-Otto Apel and Jürgen Habermas. Apel's own work and his 'transcendental-pragmatic perspective' (1984) made an important contribution to the development and spread of pragmatist principles.

At the same time, Apel's theoretical orientation had a significant influence on Habermas' engagement with related theories. In an interview, Habermas described how he got (re)involved with philosophy of science in the 1960s and interested in pragmatism in particular:

> 'Encouraged by my friend Apel, I also studied Peirce as well as Mead and Dewey. From the outset I viewed American pragmatism as the third productive reply to Hegel, after Marx and Kierkegaard, as the radical-democratic branch of Young Hegelianism, so to speak. Ever since, I have relied on this American version of the philosophy of praxis when the problem arises of compensating for the weaknesses of Marxism with respect to democratic theory.' (Habermas 1992, 148–149)

Habermas' work is featured in Keulartz et al.'s programmatic proposal, as part of their envisioned tasks for pragmatics ethics. Specifically, the authors refer to 'discourse ethics' (Apel 1988; Habermas 1990, 1994) as an approach for examining the conditions for forming moral norms. Table 1, which is a shortened/simplified version of the original graph included in Keulartz et al.'s paper, provides an overview of the tasks suggested in their proposal.

	Product	Process
Context of justification	a) Traditional ethics	b) Discourse ethics
Context of discovery	c) Dramatic rehearsal	d) Conflict management

Table 1: Keulartz et al.'s 'Tasks for a Pragmatist Ethics' (2004, 19; simplified table).

Drawing on Caspary (2000, 153ff.), the authors elaborate that pragmatism is as much interested in ongoing techno-moral developments and negotiations of related values as in socio-technological outcomes and their significance. These two domains are respectively described as *process-* and *product-focused* perspective. The notions of *context of justification* and *context of discovery* further specify the role and tasks of pragmatist ethics. The former refers to the critique of arguments, mobilised values, and justifications brought forward with regards to a certain product or process. The latter stresses the role of pragmatist ethics which goes beyond analytical involvements. In addition, it creates new conceptual or terminological frameworks, and facilitates societal negotiations and conflicts.

The grid, according to Keulartz et al. (2004), functions as an overview of possible tasks in pragmatist ethics, but not as a 'checklist' to be covered to an equal extent under all circumstances (2004, 18). In this book, I focus on an analysis of (a) communicative negotiations and (b) discourse ethics ; Habermas 2001 [1993], 1990). The main reason for this is that we do not have sufficient insights yet into which moral problems are negotiated by whom in this field. Moreover, it is not clear how the institutional, often corporate embedding of big data interrelates with possibilities for public debate and decision making. Therefore, before suggesting or exploring approaches to conflict management, it seems sensible to address the conditions for such approaches.[21]

Discourse ethics is a 'discourse theory of morality' (Habermas 2001, vii). It is rooted in two main normative principles, the 'principle of discourse' (D) and the 'principle of universalisation' (U).[22] Both should be understood as counterfactual idealizations meant to guide (moral) reasoning (Rehg 2015, 30). The first principle (D) states that valid norms are those that meet, de facto or hypothetically, the approval of all affected individuals. The second principle (U) proposes that valid moral norms are formed under conditions ensuring that individuals affected by their ramifications can autonomously accept these norms. Deliberations and efforts concerning discourse ethics are dedicated to ensuring democratic, fair processes of public debate, deliberation and decision making.

In this sense, discourse ethics aims '[…] to develop procedures and institutions that guarantee equal access to public deliberation and fair representation of all relevant arguments' (Keulartz et al. 2004, 19). While highlighting the conditions and presuppositions of moral discourses, Habermas' theory likewise shifts emphasis to the formative power and social significance of language. It pays attention to how communication constructs meanings, structures thought and socialising processes. Social meanings here are produced and negotiated in 'communal determination through public processes of interpretation' (Cronin 2001, xiii).

Habermas noted critically that prior theories concerning aspects of discourse ethics were faulty, because of their tendency '[...] to collapse *rules, contents, presuppositions* of argumentation and in addition confused all of these with moral principles' (1990, 93–94). Related to this, he also emphasises that '[...] (U) merely expresses the *normative content of a procedure of discursive will formation* and must thus be strictly distinguished from the substantive content of argumentation' (122, emphasis added). The notion of discourse ethics therefore needs to be understood in the context of Habermas' 'theory of communicative action' (see e.g. 1981, 1987). According to Habermas human communication poses validity claims regarding truth, (normative) rightness, and sincerity (often translated as authenticity).

In most daily domains, interaction through communication is grounded in an implicit consensus regarding commonly accepted knowledge and norms, as expressed in validity claims. These undisrupted forms of interaction are defined as 'communicative action'. In contexts, however, where dissensus emerges, participants involved in this 'disrupted communicative action' need to move toward a level of argumentative discourse: Habermas defines this '[…] practical discourse as a reflective continuation of communicative interaction' (McCarthy 2007, xi). During such discursive negotiations, validity claims to knowledge and values – i.e. truth, (normative) rightness and authenticity/sincerity – are collectively and publicly examined.

As Mittelstadt et al. state '[c]ommunicative action requires the speaker to engage in a discourse whenever any of these validity claims are queried' (2015, 11). This quote also indicates that Habermas' discourse ethics theory is a cognitivist approach. He assumes that the validity of moral values and social norms is constructed rationally, that is, similar to the agreement on knowledge or 'facts'[23] (which, in accordance with pragmatist principles, are likewise fallible and may be subject to renegotiation). In the discursive process, these are collectively negotiated and it is then established if claims can be rationally justified as true, normatively valid, and/or sincere. According to Habermas, the main cornerstone of *moral* discourse is, however, not 'truth'. Instead, of major concern for moral reasoning are validity claims to normative rightness and how these may be negotiated in practical discourse.[24] The normative rightness of validity claims, once challenged, can only be negotiated in collective debates involving and concerning the positions of all affected actors.

It is important to note that Habermasian discourse ethics is not normative in the sense that it assesses *content as such* as morally (un-)reasonable. Instead, its normative angle is grounded in the question whether social norms arise under conditions justifying their validity. This moral theory is consequently less concerned with traditional questions of the good life or happiness, but mainly with issues of (social) justice (see also Habermas 2001 [1993], 151; Cronin 2001, xxiii). According to Habermas, valid social norms are those which ensure justice. In this sense, '[…] a norm is just or in the general interest means nothing more than that it is worthy of recognition or is valid. Justice is not something material, not a determinate 'value,' but a dimension of validity.' (Habermas 2001 [1993], 152).

Drawing on discourse ethics seems particularly appropriate and relevant, seeing that justice is likewise a core concern for critically examining the societal implications of datafication and (big) data (Taylor 2017; Heeks and Renken

2016; Johnson 2014). Based on the assumption that only certain conditions and presuppositions for the forming of social norms foster their validity and/ as justice, Habermas reasons '[…] that all voices that are at all relevant should be heard, that the best arguments available given the current state of our knowledge should be expressed, and that only the unforced force of the better argument should determine the 'yes' and 'no' responses of participants' (2001 [1993], 145; see also 35–36).[25]

Habermas defines contexts corresponding with all these presuppositions as 'ideal speech situations'. He describes this term as regrettably somewhat misleading (2001, 163–164), since it led to criticism that it could be read as hypostatization. He rejects this reading and instead suggests it as '[…] an idea that we can approximate in real contexts of argumentation' (163). As indicated in the quote above, three aspects characterise the ideal speech situations, ensuring that validity claims may be fairly assessed and (re-)evaluated.

- Actors should not be affected by any **factors of influence** which may distort their insights into an issue or lead to their subordination due to external incentives, existing or anticipated dependences or inequalities.
- All **positions** affected by negotiated norms or knowledge are pertinent to deciding whether a validity claim is (in-)valid; thus, they should be heard and involved in the discursive process.
- Only the most coherent and just **arguments** should be decisive for decisions emerging from argumentative discourse.

These presuppositions define the conditions for 'rational acceptability' (Cronin 2001, xv). An underlying assumption is that the reaching of consensus concerning a moral value or claim to knowledge is not a sufficient condition for asserting the rationality and fairness of decision making processes. Consensus as such does not yet allow for any conclusions about the validity of result. It may always turn out that what was assumed to be based on rational consensus '[…] involved ignoring or suppressing some relevant opinion or point of view, was influenced by asymmetries of power, that the language in which the issues were formulated was inappropriate, or simply that some evidence was unavailable to the participants' (Cronin 2001, xv).

The three abovementioned presuppositions can be translated into the following main implications and questions for my analysis: first, it needs to be assessed which actors are involved in and affected by big data research. As also indicated by Keulartz et al. and Habermas' principles (U and D), discourse analysis requires a stakeholder analysis. It aims to examine '[…] who has a stake in the matter in question and should consequently have a say in the debate?' (Keulartz et al. 2002, 19). It therefore also needs to be scrutinised in which ways and to what extent affected actors were heard in relevant debates and decision-making processes. Second, the conditions for the formation of arguments and public debate need to be interrogated: i.e. to what extent the 'unforced force of the better argument' (145) was indeed decisive for the choices and responses of the affected actors. In the context of big data-driven research, it seems specifically relevant to address factors which may tilt the conversation.

Third, if one wants to approximate towards the presupposition that the 'best arguments available to us given our present state of knowledge are brought to bear' (Habermas 2001 [1993], 163), it is crucial to examine which arguments have been brought forward with regards to big data research. To do so, I draw especially on Habermas' notion of validity claims, focusing on the relevance of claims to normative rightness as well as truth, since these are discursively interlinked in many of the investigated cases. Moreover, it is necessary to evaluate to what extent arguments have been incorporated in public debate and relevant decision-making processes. For my analysis, this translates into the following theoretically grounded, guiding questions:

1. What are the broader *discursive conditions* for big data-driven public health research?
   a. Which actors are affected by and involved in such research?
   b. Which factors may shape the views of affected actors and their engagement in public discourse?
2. Which *ethical arguments* have been discussed; which *validity claims* have been brought forward?

With specific regards to big data-driven research on public health surveillance, the first question, including the two sub-questions, is examined in Chapter 4. The second question is mostly addressed in Chapter 5, although both chapters indicate how these two key issues are interrelated.

It has often been argued – and might be objected at this point at the latest – that Habermas' presuppositions set the bar unrealistically high. Obviously, neither pre- nor post-big data conditions for public debate and negotiations of social norms adhere to these principles. Habermas recognised this aspect as a core issue regarding the practical implementations of his theory, and raised, among others, the question: 'How can political action be morally justified when the social conditions in which practical discourses can be carried on and moral insight can be generated and transformed do not exist but have to be created?' (2007, 210) While the suggested presuppositions are neither achieved nor achievable societal conditions, they are nevertheless useful benchmarks of orientation. They allow us to assess whether we are moving closer to or further away from conditions fostering valid social norms which are, in this case, decisive for research ethics.

Starting from the questions stated above, I therefore aim at showing to what extent practices and discourses regarding big data research move towards or further away from presuppositions key to valid social norms. The indicated questions will be especially relevant to my analysis of discursive conditions in Chapter 4 and specific projects in Chapter 5. The following Chapter 3 will also refer back to Habermasian discourse ethics, albeit more sporadically, as it is mainly meant to provide an overview of more general, ethical issues concerning big data. It particularly serves as a primer for those unfamiliar with ethical issues concerning big data and their use in research more generally.

Big Data: Ethical Debates

In their research, scientists continuously make decisions that need to balance what they can do and what is morally reasonable to do. This applies notably to innovative research at the forefront of technological developments. In research projects located at universities, and in democratic societies, such decisions are commonly not simply made by isolated individuals or research groups. Biomedical research and studies involving human subjects in particular have become increasingly regulated in this respect, with *Institutional Review Boards* (IRBs)/*Ethics Review Boards* (ERBs) and *Research Ethics Committees* (RECs) playing a decisive role.

With regards to regulatory efforts and research ethics, Hedgecoe (2016) observes:

> 'The most obvious regulatory growth has been in the bodies responsible for the oversight of research, on ethical grounds, before it is done (a process referred to here as 'prior ethical review') – for example, institutional review boards (IRBs) in the United States, Research Ethics Committees (RECs) in the UK, research ethics boards in Canada – which have become progressively more powerful, with more kinds of research falling under their remit and with greater control over the research they oversee.' (578)

These boards and committees are often established at universities, relying on peer evaluation by scholars with (ideally) expertise in respectively related fields.[26] Governmental funding agencies are especially likely to request such ethical approval, issued by institutional ethics review bodies, prior to the start of research projects. In some cases, intermediate assessments are also required. Likewise, some journals ask for confirmation of the ethical approval of a piece

---

How to cite this book chapter:
Richterich, A. 2018. *The Big Data Agenda: Data Ethics and Critical Data Studies.*
Pp. 33–51. London: University of Westminster Press.
DOI: https://doi.org/10.16997/book14.c. License: CC-BY-NC-ND 4.0

of research (which does not necessarily mean though that they demand written proof of this).

As stressed by Hedgecoe (2016, 578), biomedical research has become more regulated over the last 50 years. This field has a comparatively long tradition in establishing ethical principles. This is arguably different to the more recently emerging applications of data science and big data-driven research. While big data may allow for biomedical insights, their retrieval is not necessarily classified as an approach that falls under regulations that have been established for non-interventional/observational biomedical research.

Since emerging technologies related to big data potentially open up previously unavailable opportunities for research, ethical questions will be also (at least partly) uncharted territory (see e.g. Mittelstadt and Floridi 2016; Zwitter 2014; Swierstra and Rip 2007; Moor 2005). This matter becomes even more complicated when considering that such research does not only take place in university departments. Internet and tech corporations themselves also conduct research, circumventing forms of ethical oversight as they apply to universities (Chen 2017; Rothstein 2015).[27]

Under which conditions and how these dynamics play out in big data-driven public health research and surveillance will be explored in Chapters 4 and 5. As a broader contextualisation however, the following subchapters first examine more generally which ethical issues, values and norms have been at stake when discussing how big data is used in research. For this too, Habermas' theory of communicative action and the notion of discourse ethics is relevant. Both allow for a conceptualisation of how norms and moral values are formed.

As described in the previous chapter, this requires that communicative routines are challenged and debated, potentially re-organised or affirmed. I established that emerging technologies have a key role in triggering such dynamics: 'Emerging technologies, and the accompanying promises and concerns, can rob moral routines of their self-evident invisibility and turn them into topics for discussion, deliberation, modification, reassertion.' (Swierstra and Rip 2007, 6). Norms and values can be considered as tacit, moral assumptions guiding such routines.

One of the reasons why we have recently witnessed broader debates on rights and demands, such as privacy, transparency, security, autonomy, or self-responsibility, is that big data developments have challenged related norms. Therefore, it is relevant to introduce some of these negotiated values more generally before proceeding to more specific conditions and cases. I first provide an overview of privacy, security, transparency, and openness. These have been arguably core (conflicting) values in big data debates. They have been mobilised as justification for big data's relevance, as reasons for inherent risks, and as constraints to public access alike (Puschmann and Burgess 2013; boyd and Crawford 2012). Calls for openness and transparency are also related to the open data movement, which promotes the accessibility of data as a public good.

As I show in the next subchapter, this may conflict on the one hand with corporate data interests and on the other hand raises issues for ensuring individuals' privacy. The last three subchapters depict debates concerning informed consent, (un-)biased data, and corporate data economies. It is particularly highlighted how big data's alleged lack of biases is brought forward in ethical debates concerning the relevance of informed consent. In contrast to the common 'digital positivism' (Mosco 2015) when referring to big data, I stress the role of algorithmic biases and how these reflect the tech-corporate contexts in which large parts of big data are being created.

Privacy and Security

Privacy and security are arguably among the most extensively discussed concerns regarding big data uses.[28] As I will show further below, they are a well-established, but misleading dichotomy. Privacy denotes individuals' possibilities for defining and limiting access to personal information. This may relate to bodily practices, fo example unobserved presence in personal spaces, or to information generated based on individuals' digital traces (see e.g. Lane et al. 2014; Beresford and Stajano 2003).

Regarding individual privacy, big data critics have emphasised individuals' (lack of) control and knowledge concerning the personal information collected when using online services (Tene and Polonetsky 2012; Lupton 2014d). This aspect is also closely related to diverging opinions on individuals' responsibility to protect their privacy, and data collectors' moral liability for fair service conditions (Puschmann and Burgess 2013). While big data proponents, and corporate service providers in particular, insist that users' information remains anonymous (Hoffman 2014), critics have raised doubts about the very possibility of anonymising data of such diverse qualities on such a large scale (Ohm 2010).

In democratic societies, privacy is considered a civic right. The *right to privacy* is (implicitly or explicitly) anchored in many national constitutions (González Fuster 2014; Glenn 2003). The *protection of personal data* tends to be considered as an extension of the right to privacy. However, the Charter of Fundamental Rights of the European Union treats them separately, with Article 8 focusing on data protection, and respect for private and family life being covered in Article 7 (The European Union Agency for Fundamental Rights, n.d.).

More recently established rights, such as the *right to be forgotten*, as established in Argentina and the EU, are closely related to (although distinct from) the right to privacy. In a 2014 ruling, the *Court of Justice of the European Union* decided that '[i]ndividuals have the right – under certain conditions – to ask search engines to remove links with personal information about them' (European Commission 2014, 1-2). This has been described as a strong signal that 'privacy is not dead' and that the EU approach contrasts with US 'patch-work' privacy policies (Newman 2015, 507).

Restrictions apply to the right to be forgotten where it conflicts with major public interests. This also implies that it '[…] will always need to be balanced against other fundamental rights, such as the freedom of expression and of the media' (European Commission 2014, 2). The criticism has been made that this decision is partly left to corporations owning respective search engines, notably to market leader Google. Freedom of speech, as well as the right to safety, have been particularly underscored as rights and values countering individual privacy considerations. These balancing acts, weighing individual rights against the public interest, are also characteristic of ethical debates concerning public health surveillance.

Apart from individual privacy, big data have revived attention on the issue of 'group privacy' (Taylor, Floridi, van der Sloot 2016; Floridi 2014; Bloustein 1976). This notion implies that privacy is not merely a right which should apply to persons, but likewise to social groups. As Floridi (2014) observes, the value of privacy has been predominantly contrasted with that of (public) security: 'Two moral duties need to be reconciled: fostering human rights and improving human welfare' (Floridi 2014, 1). He opposes the assumption, however, that the latter would be a political concern regarding the public at large and the former an ethical issue concerning individuals' rights.

In the spirit of pragmatist ethics' anti-dualism, i.e. its suspicion towards dichotomies, Floridi claims that a focus on these two positions of the individual and society overall is too simplistic. Such a limited viewpoint ultimately overlooks aspects relevant to broader societal dynamics. In consequence, the ethical debate lacks consideration for certain validity claims to normative rightness. Not merely individuals, but likewise groups should be considered as holders of privacy rights. This, according to Floridi, is increasingly of importance in an era of open and big data, since individuals (especially in their role as consumers) are commonly targeted as group members.[29]

Balancing privacy and security is closely related to one of the tensions predominantly stressed in public health research and biomedical research more generally: safeguarding individual, civic rights versus public health and wellbeing as a common/public good.[30] With regards to genomics research, Hoedemaekers, Gordijn and Pijnenburg emphasise that '[a]n appeal to the common good often involves the claim that individual interests must be superseded by the common good. This is especially the case when the common good is seriously threatened' (2006, 419).

To determine when a society may be 'seriously threatened' (e.g. by a disease) is however not always as clearly discernible as for instance in the case of epidemics/pandemics: for example, when it comes to preemptive measures such as coerced vaccinations. Moreover, the response to a perceived threat depends on the respective understanding of values relevant to the 'common good' (London

2003). In this sense, conceptualising data as contribution to the common good becomes a crucial factor in justifying their means of collection. It is therefore particularly insightful and relevant to address how tech corporations take an interest in demonstrating how 'their' big data allow for insights beneficial to societies' wellbeing – with (public) health being a widely acknowledged factor in this.

Open Data

One can observe controversies around the 'trade off' between privacy (commonly depicted as an individual right) and security (commonly depicted as a value related to public welfare, public interest and the common good) vividly with regards to governmental surveillance, as well as tech-corporate support of and acquiescence in such practices (see also Chapter 2). At the same time, transparency has been mobilised in claims to the *normative rightness* of big data practices (Levy and Johns 2016).

Transparency indicates a high degree of information disclosure. It implies openness regarding features and processes: for instance academic, governmental, corporate, or even private practices. The notion is commonly linked to accountability. With the concept of *open data*, transparency has been applied to big data as such: 'Open data is data that can be freely used, re-used and redistributed by anyone – subject only, at most, to the requirement to attribute and share alike.' (Open Knowledge International. n.d.; see also Gurstein 2011). The concept applies to data which are comprehensively accessible and technically as well as legally modifiable, allowing for re-use and distribution.

Open data can be seen as a form of output-transparency. They allow for insights into the kinds of data collected by governments or research actors/ institutions, granted even to external actors who were not involved in the initial data collection process. Open data emphasise transparency and sharing as a moral duty and quality feature. While acknowledging the potential advantages of open data, authors such as Levy and Johns advise caution when it comes to such claims. They argue that certain actors may also 'weaponize the concept of data transparency' (2016, 4). The authors stress that '[…] legislative efforts that invoke the language of data transparency can sometimes function as 'Trojan Horses' designed to advance goals that have little to do with good science or good governance' (2; see also Iliadis and Russo 2016, 3ff.).

Openness and transparency have not only been applied to data as product, but also to data collection processes. In data collection – be it for research, commercial purposes, or governmental statistics – transparency regarding procedures and purposes is known to positively influence individuals' willingness to compromise on privacy (Oulasvirta et al. 2014). For quantitative research, transparency is, moreover, a crucial methodological criterion to ensure the reproducibility of results (Stodden 2014). Both aspects are challenged in most big data practices however, since the level of transparency is considerably limited.

While open data have gained in importance (World Wide Web Foundation 2016; Kitchin 2014), most corporate data are still inaccessible to civic actors – except if they are paying (advertising) customers or commissioned researchers. Access to big data is in most cases a privilege of actors affiliated with corporations or research projects (boyd and Crawford 2012; Manovich 2011). Such corporate limitations in data access are usually presented as a means for ensuring users' privacy, but have obvious economic advantages too. Data allow for insights into (potential) customers' attitudes and behaviour, ensuring an economic advantage and making these data valuable commercial assets (see also the last subchapter below). Individuals have to rely on assurances that their data are used only in limited ways. Due to this common limit on access to big data for non-corporate, external actors, such as researchers or users themselves, such actors can hardly assess claims regarding how data are anonymised, collected or utilised. In this sense, as long as certain, corporate big data are not indeed published as open data, one may claim openness regarding the processes, but the actual material itself is not transparently accessible.

As mentioned above, it is commonly argued that this lack of transparency is needed in order to safeguard customers' privacy (Puschmann and Burgess 2013; boyd and Crawford 2012). One may query though what other motives are relevant to this mobilisation of privacy, or how this influences, for example, companies' investments in data anonymisation (see also Mattioli 2014). The very possibility of anonymising certain (big) datasets has been fundamentally called into question (Ohm 2010). In light of these challenges, it seems even more worthy of discussion that such data are being collected and used in commercial contexts, among others.

Big data enforce an increased, though neither necessarily deliberate nor conscious, transparency of online users/consumers. The full extent of this transparency is only visible to those actors controlling the main data collecting platforms or gaining external access to these (Andrejevic 2014, 1681). What is ultimately collected here, are vast amounts of personal information, concerning individuals' preferences, attitudes, moods, physical features, and – as emphasised in this book – health status and health-relevant behaviour. With the advent of big data, the notion of transparency has been increasingly applied to and demanded from individuals and their practices (O'Hara 2011).

The delusive expression 'I have nothing to hide' has been popularised in a post-9/11 era when individuals globally felt that their personal integrity should stand back in favour of public welfare and safety (see also Levi and Wall 2004). In this context, similarly to Floridi (2014), Solove (2011) observes that '[…] when privacy is balanced against security, the scale is rigged so that security will win out nearly every time' (207; see also Keeler 2006). In order to weigh up these complex values though, one needs to be aware of the full implications of privacy breaches. However, considering the lack of consideration for group privacy, many aspects are still neglected in current debates and decision making processes.

While individuals may be more willing to compromise on their privacy when it comes to security and public welfare/common good, this is often not their main motive for providing and producing personal data. It has often been suggested that 'convenience' is a main factor for the easiness with which users' allow access to their personal data. This occurs in some instances in a rather condescending tone (see e.g. the quotes by *Gnip* CEO Jud Valeski in Puschmann and Burgess 2014 or McCullag 2008) or as a comparatively neutral observation (Craig and Ludloff 2011, 1 and 13). Terms such as 'convenience', or even 'ignorance', should however instead be translated into 'lack of choice' and 'built-in data trades'.

Apart from the decision to opt-in or opt-out, in most situations, users have only marginal leeway in defining which data may be collected. In order to use services such as social networking sites or effective search engines, users have to agree to their data being used by the companies owning these platforms. Opting out of these platforms likewise implies opting out of the social benefits which these offer. Not using a particular search engine may result in a lower quality of information retrieval; not being present on a popular social network may affect a persons' social embeddness. In light of the relevance of digital media for individuals' private and professional life, drawing on such services is no longer a matter of convenience and personal choice, but of societal expectations.

As Andrejevic points out, simplifying users' behaviour as a well-balanced, conscious trade of privacy in favour of convenience ignores the power/knowledge relations emerging between contemporary digital platforms and users: 'This framing of the exchange assumes people are aware of the terms of the trade-off and it construes acquiescence to pre-structured terms of access as tantamount to a ready embrace of those terms.' (Andrejevic 2014, 1682) This is related to the accessibility and intelligibility of terms of services and privacy policies, but also to the seamless embedding of data sharing in digital media use, and the lack of practical insights into its (negative) consequences (ibid.).

The compliance of many users in giving away data to access certain services stands in stark contrast to the lack of public insight into corporate big data practices: into their contemporary collection, documentation, possible ramifications and future uses. Andrejevic speaks fittingly of a 'big data divide' (2014), referring to '[…] the asymmetric relationship between those who collect, store, and mine large quantities of data, and those whom data collection targets' (1673).[31] This notion inherently rejects the often implicit assumption that users' data sharing is simply a matter of well-informed, deliberate choices. Likewise, it emphasises the non-transparency experienced by those civic actors producing big data, and the power imbalances inherent to datafication.

Data Asymmetries and Data Philanthropy

Big data are often inaccessible data, especially when it comes to those produced on commercial platforms. While open data are becoming more common for governmental, scholarly or institutional datasets (although resistance is also notable in these domains), this kind of accessibility has not yet taken off among corporations yet: 'Despite the growing acknowledgement of the benefits, we are far from having a viable and sustainable model for private sector data sharing. This is due to a number of challenges – most of which revolve around personal data privacy, and corporate market competitiveness.' (Pawelke and Tatevossian 2013)

The lack of accessibility implies that actors looking at these data from a corporate perspective (or commissioned by respective companies) can assess what kind of information is revealed about the individuals generating these data. Moreover, only those 'insiders' have insights into the possibility of anonymising information concerning users. This lack of accessibility condemns most actors and social groups in contemporary societies to speculation about the possibilities and risks of big data.

Big data function as crucial 'sense-making resources in the digital era' (Andrejevic 2014, 1675). On the one hand, they allow for the production of knowledge concerning, for example, individuals associated with a certain profile (email, social network, etc.) or IP address. On the other hand, they would also allow for a concrete assessment of ethical concerns. This is hindered, however, because big data's accessibility is not systematically granted to company-external actors in, for example, mandatory data audits. Therefore, the big data divide implies power/knowledge conditions that systematically exclude individuals from access to data which would allow them to assess the data generated by corporations, the conditions under which this is done, and how this information is used.

According to boyd and Crawford, the lack of such independent access to big data results in a problematic constellation of 'data rich' and 'data poor actors' (2012, e.g. 674). The authors are notably concerned about the ramifications for research. They argue that the limitations in accessing big, corporate data create a '[...] restricted culture of research findings.' (2012) This may lead to a bias, due to the kinds of (uncritical) questions which are being asked or due to the privilege given to leading, prestigious universities for (good publicity) collaboration. Moreover, boyd and Crawford cite a scholar who suggested '[…] that academics should not engage in research that industry 'can do better'' (ibid.). While this assessment is problematic as such, since it backs up the aforementioned asymmetries and related risks for research, it also hints at another issue. The research skills necessary for using big data can be mainly trained by company employees or commissioned scholars. The biased, unregulated accessibility of big data also raises the risk that large parts of the academic community are unable to train skills relevant to assessing these kinds of data.

In this context, one should not only speak of a big data divide, but also scrutinise the risk of data monopolies. The phrase 'big data divide' emphasises the tensions resulting from asymmetries in data access. It calls attention to the biased capacities for gaining insights into this material and assessing its implications. In addition, the term 'data monopolies' stresses that this divide not only characterises customers' lack of agency, but the market dominance of very few internet and tech corporations. Addressing practical challenges in information systems research, Avital et al. (2007) discuss the influence of 'data monopoly/ oligopoly' as a sector for data utilisation with high complexity and low (public) availability. It is '[…] populated with large companies or agencies that collect and analyze systematically large datasets for resale or other for-profit activities. (e.g., ITU, IDC, Gartner, OECD, US Census)' (Avital et al. 2007, 4).[32] To this list, one should also add leading tech companies such as Google and its parent company Alphabet Inc., Facebook and subsidiary platforms/technologies such as Instagram, Whatsapp, and Oculus VR, but also increasingly popular apps such as Snapchat (owned by Snap Inc.).[33]

Avital et al. (2007) focus on targeted attempts at collecting data to shape research processes. In contrast, big data collected by companies through search engines, social networking sites, photo sharing sites, messengers, and apps more generally are the result of complex entanglements between commercial interests, interface designs, algorithmic processes and users' indication of preferences, actions or attitudes. This information is a highly profitable asset for advertising customers and for the optimisation of internal services. Legally, these constellations are further complicated by the fact that leading internet/ tech corporations are originally based in the United States, while offering their services to users outside the US whose data are likewise collected.

Yet despite the general lack of open (big) data in the private sector, certain data are in fact available to the public. For instance, Twitter Inc. makes user data, which are publicly posted on the microblogging platform, accessible through open application programming interfaces (such as the 'search and streaming' APIs). This includes public tweets, but also favs and retweets of these short posts. Even in this case, as Burgess and Bruns point out, the conditions under which Twitter data could be used have become increasingly restrictive over time (2012; see also Van Dijck 2011). Nevertheless, thanks to the partial availability of this kind of material, Twitter has become a particularly popular subject of many research papers using or reflecting on big data.[34] This development also hints at the benefits which corporations may expect from allowing access to data, going beyond direct, economic incentives: an effect which has been described with the term 'data philanthropy'.

The idea of 'data philanthropy' suggests that there are moral incentives for sharing big data. Their public accessibility is not merely framed as a question of economic value, but as contribution to the public good. This is closely related to 'open data' or phrases such as 'data commons'. The term was mildly popularised by the *United Nations Global Pulse* initiative, a flagship project promoting the use of big data for sustainable development and humanitarian action (see also Chapter 4). During the 2011 World Economic Forum, Kirkpatrick (on behalf of *UN Global Pulse*) complained that '[…] while there is more and more of this [big] data produced every day, it is only available to the private sector, and it is only being used to boost revenues' (Kirkpatrick 2011). In consequence, the author stated, big data's potential for being employed for the public good was largely neglected. He suggested the notion of 'data philanthropy' with regards to sharing big data generated in the private sector in support of development causes, humanitarian aid, or policy development.[35]

In a blog post following up on his talk, Kirkpatrick briefly referred to economic issues ('business competition') as well as ethical concerns ('privacy of their customers') as challenges to this idea. These were also given as reasons why it was not clear in which directions data philanthropy might develop. Three years later, Pawelke and Tatevossian, also on behalf of *UN Global Pulse*, stated in a blog post on public data sharing that very few datasets are truly publicly accessible (Pawelke and Tatevossian 2013).

In 2011, Kirkpatrick mainly emphasised the sharing of private sector data with the public sector. In 2015, Vayena et al. indicated another variation in which this data sharing may take place. The authors observe that in data philanthropy '[…] public–private partnerships are formed to share data for the public good' (Vayena et al. 2015). Corporations commonly do not simply release big data, but allow for controlled access granted to selected partners. As also mentioned in the above reflections on data monopolies, an ethical issue concerns the fact that access to data, epistemic possibilities, and (scientific) knowledge production are controlled by corporations. Novel constellations in public-private big data research bring up the question to what extent studies drawing on these data inherit ethical issues pertinent to the original data collection. And what does it mean when academic research asserts the credibility of commercial big data practices by fashioning them as a key contribution to the common good? This also raises the issue that not only the practices, but also the ethics of research itself may change (see also Kalev 2016). The latter point has been especially noticeable in debates concerning informed consent.

Informed Consent

Informed consent is a moral cornerstone for research involving human subjects. Its establishment goes back to the Nuremberg Code (1947), which outlines 10 research ethics principles, the first being dedicated to informed consent. The document resulted from the 1946–1947 trials of doctors who conducted experiments with humans in Nazi concentration camps (Weindling

2001). In terms of the moral considerations regarding individuals' rights and wellbeing, informed consent is crucial for ensuring, in particular, human dignity, the respect for persons, and respect for autonomy (Rothstein and Shoben 2013, 28; Lysaught 2004; Faden and Beauchamp 1986).

There are cases, for example in large-scale epidemiological research, where data have been obtained for research from existing databases without seeking informed consent (Nyrén, Stenbeck and Grönberg 2014, 228ff.). Such decisions are, however, subject to scrutiny by ethics review boards, weighing broader public health risks against harm to individuals. The fact that big data-driven research unhesitatingly forgoes informed consent mechanisms has thus sparked ethical concern among some academics. One reason for this tendency may be that '[…] the precursor disciplines of data science – computer science, applied mathematics and statistics – have not historically considered themselves as conducting human-subjects research' (Metcalf and Crawford 2016, 2). This assumption also applies arguably to some of the biomedical, big data-driven approaches emerging in recent years.

The negligence of informed consent has been especially controversial with regards to experimental, interventional research conducted in private-public partnerships. Entanglements between consent, research ethics and corporate big data practices became obvious in a much-debated study involving Facebook data, known as the 'emotional contagion experiment'. As the authors of the original report on this study describe '[t]he experiment manipulated the extent to which people ($N = 689,003$) were exposed to emotional expressions in their News Feed.' (Kramer, Guillory, and Hancock 2014). The study included a combination of two, parallel experiments during which users were either exposed to a reduced amount of positive emotional content posted by 'friends' on their news feed, or were shown fewer posts with negative emotional content. Posts rated as containing positive or negative content were respectively withheld.

The study design was meant to test whether users' perception of certain emotions in their newsfeed would increase the likeliness of them posting similar emotional (i.e. increasingly negative or positive) content. The latter was interpreted as an expression of the users' mood. This manipulation of users' newsfeeds led to public and academic debates concerning the ethical dimensions of this study (Kleinsman and Buckley 2015; Schroeder 2014; Booth 2014). The experiment was conducted in collaboration between an employee of Facebook's *Core Data Science Team* (Kramer) and two researchers of Cornell University. The report on this study was published in the peer reviewed journal *Proceedings of the National Academy of Sciences of the United States of America* (*PNAS*). With regards to this experiment, Kahn, Vayena and Mastroianni observe that '[…] the increasing number of public-private partnerships and collaborations involving data uses and reuses will raise challenging questions about balancing privacy and data sharing, as evidenced by the Facebook example and recent calls for large-scale data philanthropy projects' (2014).

What is at stake in this study goes beyond issues of privacy: it demonstrates controversial possibilities for circumventing informed consent. An inquiry by Chambers, a cognitive neuroscientist and contributor to *The Guardian* newspaper, reveals how corporate practices had an impact on decisions concerning its ethical assessment. In an email to the *PNAS* editor responsible for approving the study's original publication and to the authors, Chambers inquired about the interpretation of informed consent in this study. Later he published a screenshot of the inquiry and the editor's reply on Twitter. In particular, he asked for the reasons why the approval of institutional review boards (IRB) was not mentioned in the article, as this is required by the journal's policies.

In response, Fiske, the editor responsible., explained the decision to approve the paper for publication: 'I was concerned about this ethical issue as well, but the authors indicated that their university IRB had approved the study, on the grounds that Facebook filters user news feeds all the time, per the user agreement. Thus, it fits everyday experiences for users, even if they do not often consider the nature of Facebook's systematic interventions.' (Chambers 2014) The answer is insightful, since the reason for giving ethical approval is directly derived from a common corporate practice. In doing so, moral values relevant to the study are inferred from Facebook's corporate rationales and algorithmic approaches to users' news feeds. It was ultimately not confirmed whether this explanation was indeed provided by an IRB, but the dynamics depicted here show a realistic risk: that corporate data practices have a defining influence on research ethics involving related data.[36]

After contradicting statements regarding the IRB approval circulated, an email exchange between Fiske and journalist LaFrance was published in *The Atlantic*:

> 'When I asked Fiske to clarify, she told me the researchers' 'revision letter said they had Cornell IRB approval as a 'pre-existing dataset' presumably from FB, who seems to have reviewed it as well in some unspecified way… Under IRB regulations, pre-existing dataset would have been approved previously and someone is just analyzing data already collected, often by someone else.' The mention of a 'pre-existing dataset' here matters because, as Fiske explained in a follow-up email, 'presumably the data already existed when they applied to Cornell IRB.' (She also notes: 'I am not second-guessing the decision.')' (LaFrance 2014)

This case highlights a grey area when it comes to informed consent in the era of big data. It still remains unregulated, and it is unclear whether users' approval of social media privacy policies is sufficient in order to morally justify using their data for research purposes (see Vayena and Gasser 2016, 25ff.; Rothstein and Shoben 2013; Ioannidis 2013).

Beyond this 'emotional contagion' experiment, big data and related research practices have been described as influential factors in recent debates concerning informed consent. In response to the recent tendency to view informed consent as counterproductive, burdensome, and obsolete, Rothstein and Shoben (2013) provide an overview of the pros and cons pertinent to informed consent. Their article discusses the issue with particular regard to concerns regarding *consent bias*[37].

The authors emphasise that, from a more practical viewpoint, the extent to which consent bias emerges has been frequently overstated. In addition, they highlight that individuals' trust in research acts as the main factor in counteracting conditions that might lead to such bias. From an ethical perspective, they state that '[t]he argument that informed consent is incompatible with modern research represents an assault on the societal values on which biomedical research is based.' (Rothstein and Shoben 2013, 34).

Commenting on Rothstein and Shoben (2013), Ioannidis (2013) likewise stresses the potentially damaging consequences of compromised research ethics for the relationship between scientists and the public (Ioannidis 2013, 41). The author argues that attempts to dismantle informed consent are less related to consent bias, but rather motivated by new research possibilities enabled by big data. While such data were not collected for particular research, '[t]he exponential growth of electronic databases, suitable software, and computational power has made it very tempting to use such data for research purposes. If so, non-consenting people may even hinder research progress and undermine the public good' (Ioannidis 2013, 40). In fact, such an accusation and statement was made by data journalist Cukier with regards to not analysing data: 'Not using data is the moral equivalent of burning books' ('Not using data' 2016).

While this is of course an exaggerated, presumably deliberately provocative proposition, it illustrates a recurring line of argumentation and trope: opposing the use of big data is equated with hindering innovation. Comparably, arguments highlighting the relevance of informed consent from an ethical perspective are accused of obstructing possibilities for research. In turn, justifying the obsolescence of informed consent is necessary in order to pave the way for research involving certain kinds of big data. This justification is, for example, approached by highlighting the downsides of informed consent, such as consent bias. Furthermore, this is substantiated by framing material as 'pre-existing data sets', as illustrated with the aforementioned 'emotional contagion' experiment.

Algorithmic Bias

Informed consent has been criticised for creating 'consent bias,' in turn suggesting that biases do not apply to big data. As already indicated with the notion of 'digital positivism' (Mosco 2015) and 'dataism' (van Dijk 2014), several authors have stressed that '[…] the ideological effect of big data is the denial of the existence of ideology and bias' (Baruh and Popescu 2014, with reference to Cohen 2013). Despite this tendency, big data create new forms of bias relating back to the (often commercial) conditions under which they have been collected.

Commercial big data are retrieved from individuals that have the necessary resources, plus the skills and an interest, to use certain digital devices and platforms. Although collected in immense quantities, big data may still represent specific populations. Because individuals included in a big data sample tend to represent only those using an expensive/innovative technical device or service, these may be e.g. on average younger or above average physically active. This leads to selection (sampling) bias, also described as population bias (Ruths and Pfeffer 2014; Sharon 2016). Such bias implies that generalising claims based on big data, typically underlined with reference to the popularity of digital devices/platforms, should be treated with caution: the more exclusive (e.g. economically or due to required skills) a technology or platform, the higher the chances for population bias. Yet, '[d]espite these sampling biases being built into platforms used for scientific studies, they are rarely corrected for (if even acknowledged).' (Ruths and Pfeffer 2014)

Since Apple's *Research Kit* was released in 2015, it has been promoted as an efficient, effective possibility for recruiting study participants and collecting data. The Kit is targeted at medical researchers, allowing them to develop apps for iPhone users. These individuals may then take part in medical studies by using respective apps, thereby providing access to data tracked with their mobile devices. Apple advertises the Kit, to users, as follows: 'ResearchKit makes it easy for you to sign up for and participate in a study by using the iPhone that's already in your pocket. You no longer have to travel to a hospital or facility to complete tasks and fill out questionnaires.' ('ResearchKit and CareKit' n.d.). Moreover, it addresses researchers with the promise that '[…] the sheer number of iPhone users around the globe means that apps built on ResearchKit can enrol participants and gather data in bigger numbers than ever'. The implications of who uses and can afford these devices receive little to no attention in this context.

In their assessment of Apple's ResearchKit, Jardine et al. (2015) point out that '[t]he potential for bias is significant' (294). For researchers, this also implies that demographic data need to be collected and possible bias accounted for. Such a 'device-related' population bias may lead to a sample of users with specific demographics. As long as demographic limitations, e.g. with regards to generalisability, are taken into account and acknowledged, these are not necessarily problematic sample features (Chan, Yu-Feng Yvonne et al. 2017). But one should not forget that demographic characteristics are just the tip of the iceberg when it comes to potential bias.

As Baruh and Popescu show, certain users may entirely opt out of using particular services due to privacy concerns. This raises the issue that big data may systematically exclude certain groups for which these concerns are characteristic. The authors highlight that the common 'notice and choice' frameworks of online platforms and their data collection:

> '[…] effectively rationalize market withdrawal for the privacy-conscious individual (the Awareness Paradox), while creating new power imbalances for the individuals that fully rely on the market-produced solutions. The withdrawal, however partial, from the market of those individuals highly intolerant of privacy violations only serves to further skew market signals by legitimizing the argument that 'digital natives' have different, laxer privacy expectations' (Baruh and Popescu 2014, 14).

This argument has two main implications: First of all, little is known about the biases inherent to particular types of big data, especially those collected through corporate services such as social networking sites. Secondly, simply assuming that big data are indeed unbiased is inherently an ethical issue, since this promotes the social values derived from – potentially biased – samples. It is also related to the fallacy that the very fact that these data exist may be used as an argument that individuals should comply with how these have been collected.

Apart from the issue that these individuals have been given very little choice (Wessels 2015; Baruh and Popescu 2015, 8), the conclusions drawn from these datasets merely refer to those users who were willing (and able) to accept data collection conditions applying to a certain web service. As stressed in the abovementioned quote by Baruh and Popescu, this has an impact on the visibility and perception of certain moral values (see also Taylor 2017). While users' compliance is framed as representative, those who are deliberately more privacy conscious, have more consequent attitudes, or experience less peer-/ work-pressure concerning their use of a certain platform, are excluded from the data used to infer this assumption.

The described scenario refers particularly to the extreme case of non-users who entirely opt out of certain services, for example to avoid negative consequences for their privacy and autonomy (see also Oudshoorn and Pinch 2003). In addition, one needs to consider biases which may be fostered by the (algorithmic) conditions of platforms on which respective data were collected. This issue has been coined 'filter bubble' by Pariser (2011). With this term, the author/activist-entrepreneur calls attention to entanglements between users' 'search and click history' on certain platforms and content which they are more likely to see in consequence. For example, depending on users' interactions with content in their Facebook newsfeed, a person is more or less likely to encounter algorithmically curated content posted by certain actors.

Pariser also argues that this may have problematic consequences for the information diversity encountered by users. Effectively, over a longer time period, users run the risk of interacting with an online 'echo chamber': an environment in which their political views, values, or emotions are more likely to be confirmed than opposed, and potentially more likely to be reinforced than reconsidered. This raises the issue of individuals receiving rather one-sided information on, for example, political events, as has been argued with regards to developments such as Donald Trump's election as US president, as well as the outcome of the referendum concerning the U.K.'s withdrawal from the European Union (Jackson 2017). In light of study designs such as the abovementioned emotional contagion experiment, it seems especially precarious that actors who are not affiliated with corporate data collecting entities such as Facebook may not receive unmediated insights into how this possibility is used and to what extent it may influence users' perceptions.[38]

These are crucial deliberations for thinking about individuals' (lack of) possibilities to access diverse content online, and how social media may contribute to the formation of opinions, decision making, and discursive participation. At the same time, they are also relevant for evaluating the data being produced under such conditions (Bozdag 2013). Because users are more likely to encounter certain content, it is also more likely that they will interact with this content. These interactions are documented and translated into data which are then potentially used as a basis for various analytical processes, e.g. to instruct corporate decisions or to conduct research (or both).

However, since algorithms influence what kind of content users may interact with, and impact the data produced, this also increases the likelihood of systematic biases (see also Baeza-Yates 2016). Scholars in software and algorithm studies have long been vocal on the point that the agency of such non-material technological factors needs to be accounted for (see e.g. Manovich 2013, 2011; Friedman and Nissenbaum 1996). These deliberations are likewise relevant for the data resulting from the interplay between users, algorithms, software, platforms and their potentially corporate providers. These kinds of bias are particularly challenging, since the relevant algorithms are commonly difficult to access, in part due to proprietary claims and their being in a constant state of (commercially driven) flux.

As Rothstein and Shoben emphasise in their abovementioned reflections on consent bias, bias is an inherent part of research and altogether unavoidable (2013, 34). It seems crucial to show that, and how, this also applies to big data, since arguments for its epistemic superiority have also been brought forward in order to undermine previous research values. As the authors aptly argue, it is not the realisation of consent bias which is new, rather '[…] what is new is the claim that it constitutes a justification for dispensing with informed consent in one or more types of research' (2013: 34).

Characteristically, these 'types of research' involve big data coming along with their own – commonly downplayed – biases. As shown above, such attempts at mitigating the relevance of informed consent ignore that it goes beyond matters of physical integrity, but aims at safeguarding personal dignity and autonomy. What has been described as 'digital positivism' by Mosco (2015) has a crucial discursive function in this context: it manifests itself in claims for allegedly overcoming biases pertaining to more traditional data collection. But big data in fact introduce a complex entanglement of novel human-algorithmic biases.

In certain cases, for example, data generated on social networking sites such as Facebook or Google web search logs, the corporate interests in creating data are the main sources of bias, because they are decisive for the implemented algorithms. Therefore, a brief overview of the role of these interests will be covered in the final subchapter. Undeniably, this is a point which could be covered far more extensively, but it is not the aim of the following sections to provide a detailed evaluation of different data-driven business models. Instead, they are meant to examine some of the commercially grounded values related to big data in relation to research and public health information.

Data Economies

Even before the emergence of so-called 'Web 2.0' services allowing users' to create, publish and exchange content without having to rely on intermediaries, scholars had raised the issue of 'free digital labour' (Terranova 2000; see also Trebor 2012). In the late 1990s, online services such as the message boards and chats provided by America Online (AOL) involved users as 'community leaders' and administrators who monitored and maintained the quality of content and conversations.[39] Back then, users were charged based on an hourly rate (approx. 3,50€/hour in the early 1990s; see Margonelli 1999). Therefore, users' unpaid, affective labour as leaders/admins contributed to the profit generated by the company, since these volunteers maintained content in a way which made it more attractive for other, paying users to access AOL. Similarly, customers of more recent services and social networking sites are crucial for creating commercial value, since they generate content which incentivises others to access a platform. This tendency has been hailed as 'digital prosumption' in business contexts (see e.g. Tapscott 1996) and was later on more critically described as a form of free labour (Terranova 2000; see also Fuchs 2011; Ritzer and Jurgenson 2010).[40]

Moreover, users act as a target audience for advertisements and additional services offered by tech corporations and their business partners. Users' interactions with each other and with encountered content are crucial for determining what kind of content they will be offered. For instance, as Fuchs summarises: 'Facebook's users create data whenever they are online that refers to their profiles and online behaviour. This data is sold to Facebook's advertising clients who are enabled to present targeted advertisements on users' profiles.' (2014, 14)

These kinds of targeted advertisements, and more generally content which is likely to facilitate users' attention and contributions, are not limited to the platform through which certain data have been generated. Instead, as Gerlitz and Helmond (2013) show, they take place in complex entanglements between various platforms and services. The integration of various social buttons and the Open Graph protocol foster 'back end connectivity' between platforms. What the authors describe as a 'Like economy' creates an online environment held together by social buttons: it combines decentralised data collection with recentralised analytics and economic valorisation (see Gerlitz and Helmond 2013, 1361).

While the data generated by users are crucial assets for technology corporations and their networked platforms, it has been critically discussed that corporations' data practices should be regulated more clearly. Common concerns pertain to privacy and the need for current legal frameworks to catch up with technological developments (Crawford and Schultz 2014; Andrejevic and Gates 2014; Tene and Polonetsky 2012). Given the criticism around corporate uses of big data, research involving these data likewise becomes potentially subjected to these concerns.

Big data-driven studies may not only inherit the biases fostered in commercial, online settings, but also involve complex interdependencies between research ethics, data access, corporate practices and norms (see also Zimmer 2010). One may argue that users' acceptance of platforms' use policies is sufficient to justify the negligence of informed consent, especially in light of citizens' proclaimed 'duty to participate' when it comes to ensuring societies' overall wellbeing (Bialobrzeski, Ried and Dabrock 2012)[41]. However, one should not conflate a person's deliberate participation in certain public health measures with their inevitable and involuntary generation of personal, digital data which have not been collected in line with considerations for the public good in the first place.

This is another context in which Lupton's (2014d) reflections on the interplay between digital prosumption and the means by which users are addressed online with regards to personal and public health are relevant. The author argues that the commercial ideal of digitally engaged individuals has facilitated a 'digital patient experience economy' in which individuals' willingness to provide data on diseases and treatments has become morally valorised and even monetised (Lupton 2014d). This observation applies notably to patient experience and opinion websites, which require contributions from users. These developments are also reflected in a broader tendency to assume and morally expect users' readiness to contribute personal information in the form of big data for the (alleged) public good.

In conclusion, the broader issues and debates outlined in this chapter provide an overview of norms and values relevant to the use of big data, particularly in research. I have shown how privacy and security have been mobilised as a misleading dichotomy. Moreover, while privacy has been a major concern regarding big data practices, it was likewise used to justify the limited transparency on the part of actors involved in big data collection. The latter issue also points to described data asymmetries. Coming back to Habermas' idea of validity claims, such aspects are relevant to negotiations involving claims to normative rightness as moral deliberations for balancing society's overall wellbeing and individuals' civic rights.

In addition though, validity claims to truth appear to play an important role, especially considering big data's alleged epistemic superiority and effectiveness. I have illustrated this with regards to informed consent and the issue of 'consent bias'. Arguments concerning informed consent as a source of bias act as validity claims asserting truth. In consequence however, it was stressed that these arguments also have normative implications and an impact on ethical deliberations. The alleged potential of big data to generate less biased results has been advanced as an argument challenging the reasonableness of informed consent. This ignores the fact that informed consent is a priori rooted in moral values such as autonomy and personal dignity. But just as importantly, what is neglected in these attempts at justifying the methodologies and ethics of big data-driven research are insights into the biases characteristic of big data.

This chapter therefore also demonstrates that validity claims grounded in truth and normative rightness are complexly interrelated in the discourse concerning big data-driven research and its ethics. Big data's 'digital positivism' (Mosco 2015) and claims for their epistemic superiority are ultimately highly normative. They are therefore implicated in ethical debates, especially when it comes to weighing civic, individual rights and societies' overall wellbeing. The crucial institutional and discursive conditions for such processes in the field of big data-driven health research will be explored in the following two chapters.

CHAPTER 4

Big Data in Biomedical Research

Biomedical research comprises basic science/bench research as well as clinical research. It involves disciplines such as epidemiology, diagnostics, clinical trials, therapy development and pathogenesis (Nederbragt 2000). Studies in these fields aim to enhance the scientific knowledge and understanding of (public) health and diseases. Key objectives are the development of effective treatments and thus the improvement of healthcare.

Biomedical research has for a long time involved large datasets. However, big data and novel analytics approaches have been increasingly emphasised as significant trends (see also Parry and Greenhough 2018, 107ff.). Big data-driven research projects draw on data retrieved from, for instance, social networking sites, health and fitness apps, search engines or news aggregators. Critical factors for this biomedical 'big data revolution' are technological innovation, the popularisation of personal, mobile computing devices, and increasingly ubiquitous datafication (Margolis et al. 2014; Costa 2014; Howe et al. 2008).

In this chapter, I outline the discursive conditions for such biomedical big data-driven research, especially in the field of digital public health surveillance. To recapitulate, I derived two main, analytic questions from previous research in critical data studies (CDS), pragmatist ethics, and Habermas' deliberations on discourse ethics in particular:

1　What are the broader discursive conditions for big data-driven public health research?
   a. Which actors are affected by and involved in such research?
   b. Which factors may shape the views of affected actors and their engagement in public discourse?
2. Which ethical arguments have been discussed; which validity claims have been brought forward?

How to cite this book chapter:
Richterich, A. 2018. *The Big Data Agenda: Data Ethics and Critical Data Studies.*
　　Pp. 53–69. London: University of Westminster Press.
　　DOI: https://doi.org/10.16997/book14.d. License: CC-BY-NC-ND 4.0

The first question, including the two sub-questions, is predominantly examined in this chapter. Chapter 5 responds mainly to question 2, by analysing ethico-methodological developments, justifications and tensions concerning specific big data-driven research projects. However, I also come back to some of the issues explored below when discussing ethical arguments and specific project constellations.

The following sub-chapter starts with a reflection on what commonly classifies as biomedical data. This is followed by an overview of stakeholders affected by big data-driven public health research. Subsequently, I elaborate on some of these stakeholders in more detail, specifically those that have a notably powerful role in setting a discursive agenda for big data-driven research. Specifically, I highlight the role of (inter-)national grant schemes and corporate interests, as well as (financial) support for biomedical and big data-driven research. This focus takes into account that certain (f)actors may a priori bias the discursive conditions for public opinion formation and debate.

Strictly Biomedical?

With regards to big data developments in biomedical research, one can differentiate, very broadly speaking, between two categories of relevant data Certain data are generated from biological sources such as human tissue and body fluids. In addition, observational data, for instance patient diagnoses, are provided by clinicians and other medical professionals, and documented in medical records. Parry and Greenhough (2018) describe these types of data as *derivative* and *descriptive* bioinformation (5ff.).

Vayena and Gasser (2016) argue that such data should be considered 'strictly biomedical', referring , among others, to 'clinical care data, laboratory data, genomic sequencing data' (20). In these cases, biological material (derivative) or observations (descriptive) are transferred into digital data. However, there is another category of 'digitally-born' data that are not extracted from encounters with patients or analyses of biomedical material. Instead, these data are generated by documenting individuals' interactions with computing devices and online platforms. While often created without being intended primarily as a contribution to understanding (public) health issues, these data have shown to carry 'serious biomedical relevance' (Vayena and Gasser 2016, 17).

According to Vayena and Gasser (2016), the category 'strictly biomedical' applies to genomics. This interdisciplinary science is concerned with sequencing and analysing genetic information, i.e. the DNA in an organism's genome. While the samples and methods of data collection may be considered more 'traditional' (even though, of course, highly advanced on a technological and scientific level), developments in sequencing technologies have led to new challenges of data storage and management.

Since the finalisation of the Human Genome Project in 2003, with its complete mapping and examination of all human genes, the amount of biological sequence data has dramatically increased. One of the main reasons is that '[s]equencing a human genome has decreased in cost from $1 million in 2007 to $1 thousand in 2012' (O'Driscoll, Daugelaite and Sleator 2013, 774). In turn, this has created a heightened need for data storage options, computing tools and analytics. At the same time, it has facilitated a commercialisation of genetics and related services such as 23andMe for which regulations were only enforced with some delay (see e.g. Harris, Wyatt, and Kelly 2013a, 2013b, 2016).

The use of 'digitally-born data' is being explored in various fields of biomedical research. For example, it has been asserted that data retrieved from social media such as Twitter may contribute to detecting adverse medication reactions (Freifeld 2014) or content which may indicate depression (Nambisan et al. 2013), as well as the geo-mapping of epidemics (Chunara 2012). The significance of such data as biomedical information is context-dependant, even more so than in the case of derivative and descriptive bioinformation. Content exchanged on social media – such as, for example, posts and status updates indicating meals or eating habits – may enable health-related insights. However, these data were collected without individuals' intention and mostly without their awareness that they may be used for biomedical research (see also Chapter 3 on 'Informed Consent'). In the first place, they were created to interact with friends, peers, or broader audiences: e.g. to display or discuss experiences, opinions, achievements etc.

In this context, Vayena and Gasser (2016) pointedly stress the need for new ethical frameworks regarding the largely unregulated use of such digitally-born data (28ff.). The authors refrain, however, from calling these data 'biomedical', since they do not regard it as bioinformation in a strict sense. Instead, they describe such data as 'non-biomedical big data of great biomedical value' (Vayena and Gasser 2016, 23). In contrast, I also speak of biomedical (big) data with regards to digitally-born data. A main reason for doing so is to account for the comparable epistemic value and significance of those data. This is also acknowledged by Vayena and Gasser when they state that '[…] although biomedical data are categorized on the basis of their source and content, big data from non-biomedical sources can be used for biomedical purposes' (2016, 26). But while the authors still make a differentiation based on biological or physical observations versus digital sources, I propose not to distinguish in this case, since this may also suggest that a priori different, potentially less strict, ethics guidelines should apply.[42]

In this chapter as well as in Chapter 5, I focus on those digitally-born data whose significance for biomedical research is currently being explored. I mainly investigate research aimed at using big data for public health surveillance/epidemiological surveillance. There are two main reasons for this choice: First, this is a crucial field for which digital health data have been employed so far. Second, due to the fast-paced technological and institutional developments in collecting and analysing health-relevant data, the ethical debate is only successively catching up with big data-driven research in this domain.

Who is Affected, Who is Involved?

A first step towards assessing the formation of social norms, according to Habermasian discourse ethics, is to identify: who is affected by certain developments, who *has* a say in related debates and/or who *should have* a say. Additionally, it is relevant which stakeholders play a part in shaping the respective development in the first place. This also gives some indication of interests that these actors may discursively pursue.

The big data ecosystem of public health research is complex, and an overview of stakeholders is inevitably a simplification. That said, Zwitters' (2014) classification of big data stakeholders, into *generators, collectors* and *utilisers*, is a useful starting point. The author differentiates between: a) natural/artificial actors, or natural phenomena that *generate* data, voluntarily or involuntarily, knowingly or unknowingly; b) actors and entities that define and control the *collection*, storage and analysis of data; and c) those *utilising* the collected data, i.e. actors and entities which may receive data from collectors for further, potentially redefined utilisation (Zwitter 2014, 3). These broader categories also apply to the field of big data-driven health research, although it appears useful to add another, potentially crosscutting category: d) entities incentivising and promoting the use of big data in research, for example by providing financial support.

Biomedical big data have implications for a broad range of professions, domains and actors. For example, during a workshop on 'Big data in health research: an EU action plan', organised by the EC's Health Directorate[43] (Directorate-General for Research and Innovation) in 2015, a long list of international experts participated. The list included '[…] bioinformaticians, computational biologists, genome scientists, drug developers, biobanking experts, experimental biologists, biostatisticians, information and communication technology (ICT) experts, public health researchers, clinicians, public policy experts, representatives of health services, patient advocacy groups, the pharmaceutical industry, and ICT companies' (Auffray 2016). One extremely heterogeneous group is notably absent, though: those individuals generating the digital data that are now complementing biomedical research (see also Metcalf and Crawford 2016).

Users who contribute to digital platforms and generate big data of biomedical relevance are not necessarily doing so in their role as patients. In contrast to most derivative and descriptive bioinformation, big data are also retrieved from users who are not consciously part of a certain health or research measure. Accordingly, those individuals whose data are fed into big data-driven research are key stakeholders. They enable big data approaches, since they are the source of the data in question. However, they rarely contribute actively to the decisions made with regards to if and how personal data are retrieved, analysed, sold, and so on. Their 'involvement' is commonly limited to the opt-in or opt-out options enforced by corporate terms of services and usage conditions. As well as those users whose data are included in retrieved data sets, non-users of respective platforms should also be considered as relevant stakeholders. Non-users may be systematically excluded from benefits that other, participating users may receive (see the example of fitness trackers in Chapter 2); or they may experience pressure to participate in the generation of digital health data as these dynamics become more common.

One should not mistake 'being affected' with consciously noticing the effects of a development. This is one of the main problems that much of big data-driven research is hesitant to foreground: the ethical and practical implications of such research are largely unclear. At the very least, individuals are exposed to uncertainties regarding how the data are used and what this might mean for them as stakeholders now and in the future (see also Zwitter 2014). As personal data are automatically retrieved on an immense scale, the implications of such approaches for users' autonomy, dignity and right to privacy need to be considered. However, this is an extremely heterogeneous group of stakeholders. It needs to be seen on a case by case basis (see chapter 5), in which specific, potentially vulnerable groups, may be affected by big data-driven research projects more concretely. This also includes how they may relate to the outcome and results of big data-driven health research, for example as beneficiary or harmed party.

In their paper on the US 'Big Data to Knowledge Initiative', which I introduce in more detail below, Margolis et al. (2014) propose that '[k]ey stakeholders in the coming biomedical big data ecosystem include data providers and users (e.g., biomedical researchers, clinicians, and citizens), data scientists, funders, publishers, and libraries' (957). Here, researchers are labelled as 'users'. The wording is telling, and points towards Zwitter's (2014) category c. In big data-driven studies, researchers tend to act as data *utilisers*. They are affected by big data developments, since they are faced with what is promoted by e.g. peers or funders as novel research opportunities. Big data in this context may be perceived or portrayed as an opportunity for innovation. But, for scientists, it might also turn into a requirement to engage with this phenomenon or into a competitive trend, channelling biomedical funding into big data-driven studies. As big data utilisers, biomedical researchers are repurposing data retrieved from social networking sites and other sources. At the same time, they shape normative discourses on why and how these data may be used in biomedical research. This may further incentivise biomedical research involving big data. The ethical discourses articulated by scientists involved in big data-driven research, as well as counterarguments where applicable, are considered in Chapter 5.

Apart from scientists encouraging or discouraging specific normative discourses, also more authoritative institutions come into play in this respect.

Stakeholders representing (inter-)governmental funding programmes and grant schemes, such as Horizon 2020 for the EU or the US National Institutes of Health (NHI) programmes, have also taken an interest in big data-driven research. Big data are not only a development promising research innovation and improved healthcare, but also a way to reduce (healthcare) costs. Funding bodies and institutions are important stakeholders to consider, because they are decisive for the discursive governance of research. They set broader research agendas and appear as expressly influential stakeholders shaping discursive conditions. Therefore, this point will be covered more extensively in the next sub-chapter.

Instead of or besides derivative and descriptive bioinformation, biomedical researchers in big data-driven projects draw on data collected by stakeholders such as global internet and tech corporations. As big data collectors, the latter are key stakeholders, since they have come to be decisive gatekeepers for data access and analytics expertise. Corporate data collectors and scientific data utilisers are both discursively powerful groups. Yet (inter-)dependencies between these two may notably affect researchers' agency, in their role as big data utilisers, and their integrity and expert authority.

Researchers' big data practices and related ethical discourses are often inevitably linked to the data collection approaches of internet and tech corporations such as Alphabet and Google or Facebook. Such big data collectors define which data are retrieved, how these are processed and stored, and with whom they are shared. Moreover, these corporations progressively fund and support biomedical research. In this role, they add to (inter-)national grant schemes and funding provided by other industries, such as pharmaceutical companies. This engagement simultaneously incentivises research involving big data, a development which appears to be of corporate interest for multiple reasons.

Health data analytics as corporate services are an important development in this respect too. Being data-rich actors, internet and tech corporations have developed leading expertise in this field. This applies to the expertise of individuals employed at such companies, as well as data analytics and storage infrastructures. In this domain, one can observe two, interrelated trends: one is that researchers and/or public health agencies are acting explicitly as customers of tech corporations. They do not only draw on the data collected by tech corporations as outlined above, but may also make use of their data analytics services. The other trend is that tech corporations have shown an interest in biomedical data from public sources, since these can support them in developing and maintaining health related services.

The triple role of data collector, service provider and funding body is a defining feature of internet/tech corporations. It puts these stakeholders in a powerful position, with regards to biomedical big data generators and utilisers alike. Therefore, this aspect will be covered in greater detail in the sub-chapter after next. First, though, I expand on the role of (inter-)governmental funding schemes raised above.[44]

Funding Big Data-Driven Health Research

Due to the rising size and complexity of biomedical datasets, as well as the digital origins of certain data, computer/data science expertise has become more and more important for biomedical research. Emerging technosciences such as *bioinformatics* and *biocomputing* refer to interdisciplinary research approaches. They merge data science, computing and biomedical expertise. Scholars in the interdisciplinary research field of bioinformatics, for example, create platforms, software and algorithms for biomedical data analytics, knowledge production and utilisation (Luscombe, Greenbaum, and Gerstein 2001).

The emergence of such intersections between life/health sciences and computing is also linked to the tendency that contemporary funding schemes require technology development and private-public partnerships (see e.g. 'Information and Communication Technologies in Horizon 2020' 2015). Technological output such as software or hardware prototypes and applications is increasingly decisive for various national and transnational grants. This applies also and particularly to research on and with biomedical big data.

In 2012, the United States National Institutes of Health (NIH) launched a major data science initiative, called 'Data Science at NHI'. This involved creating a new position called Associate Director for Data Science, currently [January 2018] held by Philip Bourne, a computer scientists specialising in health research. Moreover, it established a new funding scheme called 'Big Data to Knowledge' (BD2K). The programme's main aim is to explore how biomedical big data may contribute to understanding and improving human health and fighting diseases (Data Science at NIH 2016).[45] The programme is divided into four main clusters: centres of excellence for big data computing (11 centres in 2017); resource indexing; enhancing training; and targeted software development. The latter framework provides funding for projects working towards software solutions for big data applications in health research.

The European Commission (EC) too displays a clear interest and mounting investments in big data developments. In 2014, the EC published an initial communication document titled 'Towards a Thriving Data-Driven Economy' (COM 442 final 2014). The document highlights the economic potential of big data in areas such as health, food security, climate, resource efficiency, energy, intelligent transport systems and smart cities. Stating that 'Europe cannot afford to miss' (COM 442 final 2014, 2) these opportunities, the document warns that European big data utilisation and related technologies lag behind projects established in the US. Three years later, in January 2017, a follow-up communication was released: 'Building a European Data Economy' (COM 9 final 2017) One of the aims declared in this document is to '[…] develop enabling technologies, underlying infrastructures and skills, particularly to the benefit of SMEs [small and medium enterprises]' (COM 9 final 2017, 3).

On the EC website, this big data strategy is also presented by posing questions such as: 'What can big data do for you?' Under this point/question, the first aspect mentioned is 'Healthcare: enhancing diagnosis and treatment while preserving privacy'. This emphasis indicates that big data are seen as important development in healthcare, but also that healthcare is showcased as an example of how individuals can benefit from big data. Building on these focal points, the EC provides targeted funding possibilities such as the call 'Big data supporting Public Health Policies' (SC1-PM-18. 2016) which is part of the programme Health, demographic change and well-being.

Projects like Big Data Europe, which involves a big data health pilot, also received funding from grant schemes such as 'Content technologies and information management: ICT for digital content, cultural and creative industries' (BigDataEurope 2016). Such trends relate back to the EC's *Digital Agenda for Europe* (*DAE*) (a 10-year strategy development running from 2010 until 2020) and its priority 'eHealth and Ageing'. The *DAE* aims at enhancing the EU's economic growth by investing in digital technologies. Complementing national and EU-wide efforts, it also entails endeavours for enhanced global cooperation concerning digital health data and related technologies ('EU and US strengthen collaboration' 2016). Moreover, biomedical big data funding initiatives have been set up by various governments in Europe (see e.g. Research Councils UK n.d.; Bundesministerium für Bildung und Forschung n.d.).

The World Health Organisation (WHO), as a United Nations (UN) agency, likewise takes an interest in the use of big data for health research, disease monitoring and prevention. Stressing that this development opens up new possibilities and challenges, the WHO's eHealth programme states: 'Beyond traditional sources of data generated from health care and public health activities, we now have the ability to capture data for health through sensors, wearables and monitors of all kinds' ('The health data ecosystem' n.d.). With regards to big data utilisation for public health and humanitarian action, the *WHO* collaborates closely with the UN Global Pulse initiative (see also chapter 3 on data philanthropy).

Global Pulse's main objectives are the promotion and practical exploration of big data use for humanitarian action and developments, notably through public-private partnerships (see 'United Nations Global Pulse: About' n.d.). It is organised as a network of so-called 'innovation labs': with a headquarter in New York and two centres in Jakarta (Indonesia) and Kampala (Uganda). These labs develop big data-driven research projects, applications and platforms which are closely connected to local communities in the respective area and country. Among other factors, Global Pulse was inspired by NGO research initiatives such as Global Viral (which is linked to the commercial epidemic risks analytics services offered by Metabiota Inc.), the Ushaidi crisis mapping platform, and Google Flu Trends (see UN Global Pulse 2012, 2).

This overview indicates that the 'big data agenda' (Parry and Greenhough 2018, 108), in these cases the promotion of big data's use for health research, is not simply a bottom-up development stirred by individual researchers. Instead, the trend towards big data-driven health research is incentivised by authoritative institutions and actors, also in the role of funding bodies. It could be argued of course that most of these initiatives claim to go back to democratic processes, consulting experts and other stakeholders (Auffray et al. 2016). However, these consultations tend to privilege renowned experts and, to a lesser extent, patient advocacy groups, rather than directly involving actors who are affected by big data practices because they are made part of the data generation process.

Discursively, what is accentuated in (inter-)national funding schemes and policy documents is big data's impact on economic competitiveness, innovation and societal wellbeing. Considerably less emphasis is put on potential risks and uncertainties, although some improvement has been noticeable during the last two years. Thus, as stakeholders, these institutions also contribute to establishing big data as a field of interest for scientific research. The economic advantages, innovation potential and health benefits, alleged in respective grant schemes or policy documents, are authoritatively promoted as research rationales.

The Role of Tech Philanthrocapitalism

Apart from national and intergovernmental initiatives, private and corporate funding opportunities also play a role. Historically, this is of course by no means a new development in (biomedical) research. For example, in the US it was only in the 1940s that '[t]he national shift from primarily philanthropic to governmental funding took place as the National Institutes of Health (NIH) became the main vehicle for research' (Brandt and Gardner 2013, 27; see also Cooter and Pickstone 2013). In Europe, philanthropic organisations such as the (American) Rockefeller Foundation were very influential, notably in the context of World Wars I and II (Weindling 1993).[46] What is new however, is the peculiar role of internet and tech corporations. These companies have very specific interests and agendas, especially with regards to how their products may feature in contemporary research and in relation to public policies. Moreover, they invest in the development of health technologies considered auspicious additions to their product portfolio. In 2016 and 2017, for example, increasing venture capitalist and private equity funding was reported for digital health technologies (see e.g. Silicon Valley Bank 2017; Mercom 2016).

It has been noted that tech corporations increasingly receive public funding. Regarding privately held or mediated databases, Sharon (2016) observes that '[…] public money is channelled, indirectly or directly, to their development, as has been the case with 23andMe, which recently secured a US$1.4 million research grant from the NIH to expand its database, and with recent National Cancer Institute funding of Google and Amazon run genome clouds' (Sharon 2016, 569). These developments are part of the emerging data, analytics, skills and infrastructure asymmetries depicted in Chapter 3. It is important to be aware of money and data not only flowing from tech corporations to (public) research institutions, but also vice versa. Since I mainly focus on studies conducted by academics at universities, however, the following sections describe investments and funding provided by internet/tech corporations for such research projects.

More generally, it has been argued that '[…] a transition from public to private sector funding has already taken place in some domains of the sciences' (Inverso, Boualam and Mahoney 2017, 54). One of these domains is biomedical research. A report by the American Association for the Advancement of Science shows that while federal government funding is still the main source for research, 'industry has caught up' (Hourihan and Parkes 2016, 6). A well-known issue in this context is that private funding tends to privilege research that promises to deliver short-term results and product development (ibid.). While private companies spend 80% of their research and development investments on development, only 20% go into basic and applied research, a ratio which is reversed for federal nondefense agencies in the US.

Even before the big data hype, in the early 2000s scholars observed that in the US, industry influence on biomedical research had dramatically risen within two decades (Bekelman, Li and Gross 2003). Based on an analysis of articles examining 1140 biomedical studies, Bekelman, Li and Gross (2003) showed that statistically '[…] industry-sponsored studies were significantly more likely to reach conclusions that were favourable to the sponsor than were nonindustry studies' (463). From an ethical perspective, the authors problematise conflicts of interests emerging from entanglements between researchers and industry sponsors.

These entanglements have a bearing on the results that certain research may generate. Furthermore, considering industry's tendency to sponsor development-driven research, this sways the type of studies being conducted. Given such earlier insights, we should carefully scrutinise how internet and tech corporations support and fund scientific research. Financial or in-kind support is commonly made in domains that are relevant to their economic, tech-political interests and their favourable public perception.

With regards to Google, a 2017 report published by the Google Transparency Project, an initiative of the US Campaign for Accountability, comes to the conclusion that: 'Google has exercised an increasingly pernicious influence on academic research, paying millions of dollars each year to academics and scholars who produce papers that support its business and policy goals' (Google Transparency Project 2017). The report highlights among other things that between 2005 and 2017, 329 research papers dealing with public policy issues in the interest of Google were funded by the corporation. Moreover, corporations such as Alphabet, as Google's parent company, are heavily investing in biotechnology start-ups.

In 2009, Alphabet launched Google Venture (GV) as its venture capital arm. Since then, GV has invested, for instance, in 23andMe[47], Doctor on Demand,

and Flatiron, a company developing cloud-based services for oncological (cancer research and care) data. Four years earlier, in 2005, Google started its charitable offshoot Google.org. In 2017, it was stated on the website of this Google branch that it annually donates '$100,000,000 in grants, 200,000 hours, $1 billion in products'. Investments and grants are particularly targeted at projects exploring how new technologies and digital data can be used to tackle societal and ecological challenges. Various Google-sponsored tech challenges/competitions worldwide complement these efforts.

Since 2016, 'Crisis Response' has been one of Google's declared focal points, next to 'Disabilities', 'Education and Digital Skills', and 'Racial Justice'. The crisis response team was already formed in 2010, in reaction to the 2010 Haiti earthquake and the ensuing humanitarian crisis. It provides services such as Google Public Alerts, Google Person Finder, and Google Crisis Map.[48] In February 2017, Google.org specifically highlighted its efforts in 'Fighting the Zika Virus' and 'Fighting Ebola'. From 2006 until 2009, Google.org was led by Larry Brilliant. Before his appointment, the physician and epidemiologist had been involved in various enterprises, ranging from research for the WHO to co-creating the early online community The Well as well as the health-focused Seva Foundation.

After leaving Google.org in 2009, Brilliant joined the Skoll Global Threats Fund (SGTF) as managing director. The SGTF is part of the Skoll Foundation (SF), an NGO initiated by eBay founder Jeff Skoll in support of 'social entrepreneurship'. It maintains the website endingpandemics.org which describes itself as a 'community of practice' aimed at accelerating the detection, verification, and reporting of disease outbreaks globally. Similarly to the SF, the Bill and Melinda Gates Foundation, with an endowment of $44.3 billion, proposes that '[w]e can save lives by delivering the latest in science and technology to those with the greatest needs'.[49]

Not only technologies, but also the funding enabled by profitable tech corporations has been styled as an important contribution to research and healthcare. In 2016, a philanthropic investment of Mark Zuckerberg and his wife Priscilla Chan was however rather controversially discussed, at least in San Francisco. After receiving a donation of $75 million from the couple, the San Francisco General Hospital and Trauma Center (where Chan was trained as paediatrician) was renamed into the 'Priscilla and Mark Zuckerberg San Francisco General Hospital and Trauma Center' ('Mark Zuckerberg and Priscilla Chan give $75 million' 2015). The decision to rename the hospital triggered criticism from some, because it was said to ignore the continuous input of taxpayers, as well as the alarming impact of Silicon Valley on San Francisco (Heilig 2015; Cuttler 2015).

Apart from such donations, less is known about Facebook's role and interest in health research applications. Information on this has been largely speculative, partly because only few official statements are provided on the company's interests in this domain. In 2013, a report by *Reuters* suggested that the company was interested in establishing patient support websites such as PatientsLikeMe, as well as health and lifestyle monitoring applications involving wearable technologies (Farr and Oreskovic 2013). This initiative has not, however, materialised so far. Yet, Facebook often highlights its relevance as catalyst and enabler of health- relevant and humanitarian initiatives. This applies, for instance, to a status feature through which users can identify themselves as organ donors, and to 'Community Help' and 'Safety Check'. The latter are features allowing users to ask for support from others or indicate that they are safe, for example in areas hit by natural disasters.

Chan and Zuckerberg recently revealed the new health focus of The Chan Zuckerberg Initiative. This limited liability company (LLC) was founded in December 2015. After initially mainly investing in education and software training, The Chan Zuckerberg Initiative launched its science programme in September 2016. On behalf of Chan and Zuckerberg, it was declared that the programme would help 'cure, prevent or manage all diseases in our [Chan and Zuckerberg's] children's lifetime' (see also Heath 2016).

An important part of this science programme is the Chan Zuckerberg Biohub. The programme provides funds for this centre, which comprises (medical) researchers and engineers from Berkeley, University of California; University of California San Francisco; and Stanford University. In February 2017, the two main research projects were the 'Infectious Disease Initiative' and the 'Cell Atlas'. The Chan Zuckerberg Biohub, its funding structure, and its involvement of researchers are an example for emerging entanglements between university research on (public) health and tech corporations. The funding available to the 47 researchers part of the hub is unrestricted.

Zuckerberg is not the only Facebook founder investing in philanthrocapitalism. Also in 2017, the venture capital firm B Capital Group, co-initiated by Eduardo Saverin (co-founder of Facebook), invested in the technology start-up CXA group. Its declared aim was to '[t]ransform your current healthcare spending into a benefits and wellness program where your employees choose their own path to good health'. Already in 2011, another Facebook co-founder, Dustin Moskovitz, initiated the private foundation Good Ventures, together with his wife Cari Tuna. Good Ventures invests in domains such as biosecurity and pandemic preparedness, as well as global health and development.

While this is not an all-encompassing overview of corporate, philanthropically framed investments in the public health sector, it allows for initial insights into entanglements between internet and tech giants such as Alphabet and Facebook and contemporary research. More generally, since the 'Giving Pledge Campaign' was initiated by Bill Gates and Warren Buffett in June 2010[50], there has been an increase in diverse, tech philanthrocapitalist initiatives. While one may intuitively deem that philanthropic investments as such should not be seen as a problematic development, these practices raise considerable economic and ethical issues and contradictions. The Chan Zuckerberg Initiative has been described as a poster child of *philanthrocapitalism* (Cassidy 2015), a term which has turned out to be an effective euphemism for a form of 'disruptive philanthropy' (Horvath and Powell 2016, 89).

Horvath and Powell (2016) argue that disruptive, corporate philanthropy bypasses democratic control over spending in domains significant to societies' wellbeing and public good. Relating this back to Habermas' deliberations on discourse ethics, this also implies that critical public debate on such issues is largely irrelevant for these corporate decision-making processes that are not overseen by institutions embedded in democratic processes. Three main, interrelated problems should be considered here: first, emerging dependencies between corporate actors, health researchers and public health institutions; second, the tendency that large sums of otherwise taxable money are invested into philanthropically framed projects; third, the influence which corporate actors exert on content choices and developments concerning health relevant research.

With regards to Google funding, it was observed that '[t]he company benefits from good PR while redirecting money into charitable investments of its choice when, if that money were taxed, it would go toward government programs that, in theory at least, were arrived at democratically' (Alba 2016). The work of Horvath and Powell (2016) is highly insightful in this regard, since they examine how the rise of corporate, philanthropic activity is linked to the decline of democracy (89; see also Reich, Cordelli, and Bernholz 2016). According to the authors, approaches to destructive philanthropy are characterised by three key features: 1) They attempt to change the conversation and influence how societies evaluate the relevance of current challenges and possible solutions. 2) They are built on competitive values. 3) They explore new models for funding public goods. With regards to the intersection of public health research and corporate big data, these are relevant considerations. Horvath and Powell (2016) illustrate aptly how efforts in destructive philanthropy shape what is seen as societal issues, and which methods are considered appropriate for addressing respective problems (see 89ff.).

These strategies stand in stark contrast to Habermasian principles for valid social norms, notably the requirement that persons should make assessments and decisions based on the force of the better argument. Given that powerful stakeholders such as leading internet and tech corporations are shaping relevant discourses, the basis for public debate appears troubled. It is also of concern that such corporate shaping of discourses occurs conspicuously by mobilising the credibility of scientific research. Tech/internet corporations' discursive and financial engagement at the intersection of technology and biomedical research raises the question how this may shape the public perception of big data.

Furthermore, notably in the US, novel, corporate funding mechanisms influence ethics review procedures and requirements. Rothstein (2015) depicts some of the practical consequences for big data-driven health research:

'Of immediate concern is that the use of personal information linked to health or, even worse, the intentional manipulation of behavior, is not subject to traditional, federal research oversight. The reason is that these studies are not federally funded, not undertaken by an entity that has signed a federal-wide assurance, and not performed in in contemplation of an FDA [US Food and Drug Administration] submission.' (425)

As the author implies, this raises the question whether adjustments in regulations for research are needed. It also begs the question of the responsibility and capacity of corporations to ensure that funded projects are equipped with and incentivised to address ethical issues.

Tech and internet corporations take great interest in maintaining and fostering a view of (their) technologies as beneficial to scientific advancements and societal wellbeing. As part of this broader agenda, they have also come to play an influential role in heralding the benefits of big data for public health. By providing funding, data, analytics and other support, they set incentives for researchers to engage in related technoscientific explorations. In doing so, they act as important gatekeepers in defining research choices as well as implementations. This seems all the more important, since internet/tech corporations often act as crucial data and analytics providers, a tendency which is highly salient for the field of digital public health surveillance.

Digital Public Health Surveillance

'I envision a kid (in Africa) getting online and finding that there is an outbreak of cholera down the street. I envision someone in Cambodia finding out that there is leprosy across the street.' (Larry Brilliant, in Zetter 2006)

Envisioning the benefits of new technological developments is a common practice. In competitive contexts – be it for start-ups competing for venture capital or researchers competing for funding – persuasive promises emphasising the need for and benefits of a product/service/technology are indispensable. It is therefore not surprising that projects involving biomedical big data have made bold promises. As Rip observes:

'[P]romises about an emerging technology are often inflated to get a hearing. Such exaggerated promises are like confidence tricks and can be condemned on bordering at the fraudulent. But then there is the argument that because of how science and innovation are organised in our societies, scientists are almost forced to exaggerate the promise of their envisaged work in order to compete for funding and other resources.' (2013, 192/193)

This mechanism does not only apply to research. It likewise applies to corporations and their promotion of new technological developments and services, as illustrated with the above comment by Larry Brilliant. Google.org's former director ambitiously pushed and promoted its engagement in infectious disease prediction.

Epidemiology, and its sub-discipline of epidemiological/public health surveillance, has undergone significant changes since the 1980s.[51] Most recently, these are related to technological developments such as the popularisation of digital media and emerging possibilities to access and analyse vast amounts of global online user data. Epidemiological surveillance involves systematic, continuous data collection, documentation and analysis of information which reflects the current health status of a population. It aims at providing reliable information for governments, public health institutions and professionals to react adequately and quickly to potential health threats. Ideally, epidemiological surveillance enables the establishment of early warning systems for epidemic outbreaks in a geographic region or even multinational or global pandemics.

The main sources relevant to 'traditional' public health surveillance are mortality data, morbidity data (case reporting), epidemic reporting, laboratory reporting, individual case reports and epidemic field investigation. The data sources may vary however, depending on the development and standards of a country's public health services and medical facilities. Since the 1980s at the latest, computer technology and digital networks have become increasingly influential factors, not merely with regards to archiving and data analysis, but in terms of communication and exchange between relevant actors and institutions. Envisioning the 'epidemiologist of the future', Dean et al. suggested that she/he '[…] will have a computer and communications system capable of providing management information on all these phases and also capable of being connected to individual households and medical facilities to obtain additional information' (1994, 246).

The French Communicable Disease Network, with its *Réseau Sentinelles*, was a decisive pioneer in computer-aided approaches. It was one of the first systematic attempts to build a system for public health/epidemiological surveillance based on computer networks. Meanwhile, it may seem rather self-evident that the retrieved data are available online. Weekly and annual reports present intensities (ranging from 'minimal – very high activity') for 14 diseases, including 11 infectious diseases such as influenza.[52]

Similar (public) services are provided by the World Health Organisation's (WHO) 'Disease Outbreak News',[53] the 'Epidemiological Updates'[54] of the European Centre for Disease Prevention and Control (ECDC) and (only for influenza cases in Germany and during the winter season) by the Robert Koch Institute's 'Consortium Influenza'. With its *Project Global Alert and Response (GAR)*, the WHO additionally establishes a transnational surveillance and early-warning system. It aims at creating an 'integrated global alert and response system for epidemics and other public health emergencies based on strong national public health systems and capacity and an effective international system for coordinated response'.[55]

In this sense, computerisation and digitalisation have significantly affected approaches in epidemiological surveillance for decades. However, one aspect remained unchanged until the early 2000s: these were still relying on descriptive and derivative bioinformation, for example data from diagnostics or mortality rate statistics. In contrast, more recent strategies for epidemiological surveillance have utilised 'digitally-born' biomedical big data. Various terms have been coined to name these developments and linguistically 'claim' the field: infodemiology, infoveillance (Eysenbach 2002, 2006, 2009), epimining (Breton et al. 2013) and digital disease detection (Brownstein, Freifeld and Madoff. 2009).

Approaches to digital, big data-driven public health surveillance can be broadly categorised according to how the used data have been retrieved. Especially in the early 2000s, digital disease detection particularly focused on publicly available online sources and monitoring. For example, news websites were scanned for information relevant to public health developments (Zhang et al. 2009; Eysenbach 2009). With the popularisation of social media, it seemed that epidemiologists no longer had to wait for news media to publish information about potential outbreaks. Instead, they could harness digital data generated by decentralised submissions from millions of social media users worldwide (Velasco et al. 2014; Eke 2011).

Platforms like Twitter, which allow for access to (most) users' tweets through an open application programming interface, have been considered especially useful indicators of digital disease developments (Stoové and Pedrana 2014; Signorini et al. 2011). Moreover, attempts were made at combining social media and news media as sources (Chunara et al. 2012; Hay 2013). Other projects used search engine queries in order to monitor and potentially even predict infectious disease developments. The platforms *EpiSPIDER*[56] (Tolentino et al. 2007; Keller et al. 2009) and *BioCaster* (Collier et al. 2008) combined data retrieved from various online sources, such as the European Media Monitor Alerts, Twitter, reports from the US Centers for Disease Control and Prevention and the WHO. The selected information was then presented in Google Maps mashups. However, these pioneer projects seem to have been discontinued, whilst the *HealthMap* platform is still active (see Lyon et al. 2012 for a comparison of the three systems).[57]

Big data produced by queries entered into search engines have also been utilised for public health surveillance projects. In particular, studies by Eysenbach (2006), Polgreen et al. (2008) and Ginsberg et al. (2008) have explored potential approaches. The authors demonstrated that Google and Yahoo search engine queries may indicate public health developments, while they likewise point to methodological uncertainties caused by changes in users' search behaviour. Such approaches using search engine data have been described as problematic, since they are based on very selective institutional conditions for data access, and have raised questions concerning users' privacy and consent (Richterich 2016; Lupton 2014b, Chapter 5).

In this context it also seems significant that a project such as Google Flu Trends, which was initially perceived as 'poster child of big data', was discontinued as a public service after repeated criticism (Lazer et al. 2014; 2015). The platform predicted influenza intensities by analysing users' search queries and relating them to influenza surveillance data provided by bodies such as the ECDC and the US CDC. The search query data are still being collected and exchanged with selected scientists, but the project is not available as a now-casting service anymore. Instead, some indications of the data are published in Google's 'Public Data Explorer'. In light of such developments and public concerns regarding big data utilisation (Science and Technology Committee 2015; Tene and Polonetsky 2012, 2012a), ethical considerations have gradually received more attention (Mittelstadt and Floridi 2016; Vayena et al. 2015; Zimmer 2010).

While it has been discontinued as a public service, 'Google Flu Trends' is still an illustrative example which highlights how collaboration between epidemiologists and data/computer scientists facilitated research leading to a concrete technological development and public service. Some of the aforementioned authors, such as Brownstein, Freifeld, and Chunara, have also been involved in research aimed at developing digital tools and applications in digital epidemiology. For example, they created the websites and mobile applications HealthMap (which also receives funding and support from Google, Twitter, SGTF, the Bill and Melinda Gates Foundation, and Amazon) as well as FluNearYou. HealthMap draws on multiple big data sources, for example, tweets and Google News content, while FluNearYou is an example of 'participatory epidemiology' and presents submissions from registered community members.

Considering such entanglements between big data collectors and data utilisers, an analysis of individual research projects appears insightful and necessary. This chapter explored how relevant stakeholders are involved in shaping the discursive conditions for big data-driven health research. But which ethical discourses have in fact evolved under the described discursive conditions? In response, the following chapter examines which ethical arguments have been mobilised in research projects and big data-driven approaches to public health surveillance. It shows which validity claims have been brought forward. Particular attention is paid to validity claims to normative rightness, although it appears characteristic for big data-driven research discourses to interlink ethical arguments with validity claims to truth.

CHAPTER 5

Big Data-Driven Health Surveillance

The emergence of research using big data for public health surveillance is directly related to the vast, diverse data generated by individuals online. On Twitter, many users publicly post about their medical conditions, medication and habits related to self-care. By 'liking' content, Facebook users indicate their eating habits or physical (in-)activity. It is common to search the internet for information on experienced or observed diseases and symptoms. Some users sign up for online communities to exchange their personal knowledge of and struggles with illness, and some even track their physical movements and physiological signals with wearable fitness devices. Such data have come to play a role in research on public health surveillance.

When drawing on such data, especially when applying for funding and when publishing results, researchers articulate ethical arguments and validity claims contending the normative rightness of their approaches. Some of these claims will be examined in the following chapter, with specific regards to research on big data-driven public health surveillance. Important trends in this field are approaches monitoring social media, search behaviour and information access. As an alternative to mining data without users' consent, possibilities of health prosumption and participatory epidemiology are being explored.

Social media monitoring as contribution to public health surveillance. On social networking sites such as Facebook or microblogging platforms like Twitter, users post and interact with potentially health-relevant information. They may, for example, casually post about their health conditions or indicate interests and (e.g. dietary or sexual) habits which may be health-related. This sharing of information facilitates research drawing on social media data collected by tech corporations. Such research may be conducted by scientists employed at universities and (inter-)governmental institutes, and potentially in collaboration with employees of tech corporations.

How to cite this book chapter:
Richterich, A. 2018. *The Big Data Agenda: Data Ethics and Critical Data Studies.*
    Pp. 71–90. London: University of Westminster Press.
    DOI: https://doi.org/10.16997/book14.e. License: CC-BY-NC-ND 4.0

Search behaviour and information access. Due to their widespread use, search engines (most notably Google) act as main gateways to online information. Among many other things, users enter queries which may be health related and potentially allow for insights into their health conditions as well as experiences concerning, for example drugs, treatments, health providers, or physicians. Such search queries, however, are not only entered on websites which are predominantly search services. Users may also search for persons and access content related to their interests on social networking sites. Therefore, these kinds of data also play a role for the first category mentioned above. Data emerging from users' search queries can have high biomedical value in various regards. Therefore, they have been used as means for public health monitoring. Such datasets have only rarely been provided as open data, since early attempts demonstrated the difficulties of anonymisation (Zimmer 2010; Arrington 2006). Related studies have been mainly conducted by scientists employed at tech corporations, or in a few cases in public-private collaboration.

Health prosumption and participatory epidemiology. Social networking sites allow for and encourage users' participation; for example, in the form of content contributions or communal support. These forms of 'prosumption' have also facilitated the development of health platforms that engage users in ways leading to biomedical big data. In this context, research and projects have emerged which aim at developing platforms or applications needed to collect data. They are meant to create possibilities for individuals' deliberate involvement in public health surveillance as a form of 'participatory epidemiology' (Freifeld et al. 2010). Such initiatives emerged in university contexts, as part of (inter-)governmental institutions and/or businesses.

In the following subchapters, I will mainly investigate cases of social media monitoring and big data use in research on public health surveillance. I will highlight three domains: first, data retrieved from users who provide indications of physical/health conditions and behaviour, voluntarily or involuntarily, knowingly or unknowingly; secondly, data retrieved from users' interaction with social media content and features; thirdly, data retrieved, combined, and mapped based on multiple online sources. I will refer to the relevance of search queries as a data source, as well as to examples of 'participatory epidemiology'. The latter will be described in less detail though, since related approaches do not necessarily classify as big data.

High-Risk Tweets: Exposing Illness and Risk Behaviour

Especially early on, efforts in digital disease detection focused on the surveillance of influenza (e.g. Eysenbach 2006; Polgreen et al. 2008; Ginsberg et al. 2008; Signorini et al. 2011). The topical focus on influenza or influenza-like-illness (ILI) owes partly to to its widespread occurrence, but influenza is also an illness that sufferers/users tend to be comparatively open about discussing. A person who states to suffer 'from the flu' on social networking sites is relatively likely to experience sympathy (possibly also disbelief or disinterest). Individuals posting about suffering from symptoms related to their infection with the human immunodeficiency virus (HIV) may instead be subjected to stigma and discrimination.

Certain infectious diseases, such as HIV/AIDS, are known to be highly stigmatising for affected patients (Deacon 2006). This also applies to mental illnesses such as schizophrenia (Crisp et al. 2000). Affected individuals are less likely to openly and lightly post explicit information on their health condition in cases of highly stigmatised conditions. This also has implications for the accessibility of information and data regarding these diseases. It implies that certain disease indicators are reflected only implicitly and not explicitly in users' content. Despite these complicating conditions regarding big data on diseases such as HIV, studies have examined how social media can be used to monitor relevant factors. In comparison to research on big data relevant to influenza monitoring, in these cases the focus is less on articulations of symptoms, but on content indicative for risk behaviour. A difference concerning the data sources is therefore that an individual posting about or searching for information on flu symptoms is more likely to be aware what this content signifies. In comparison, a person posting about certain habits which can be classified as, for example, drug- or sex-related risk behaviour is perhaps unaware that these posts may be indicators of health risks.

As part of the BD2K funding scheme 'Targeted Software Development', several research projects explore how social networking sites could play a role in countering infectious diseases. Broadly speaking, they examine how online data may reflect users' health behaviour and conditions. Examples for projects active in 2017/18 are 'Mining the social web to monitor public health and HIV risk behaviors' (Wang et al. n.d.)[58] and 'Mining real-time social media big data to monitor HIV: Development and ethical issues' (Young et al. n.d.)[59]. Also, outside of the BD2K scheme, funding has been granted to projects such as 'Online media and structural influences on new HIV/STI Cases in the US' (Albarracin et al. n.d.)[60]. The responsible interdisciplinary research teams consist of epidemiologists, computer and data scientists, public health researchers and psychologists. Similar projects have been launched with regards to mental illness monitoring, for example 'Utilizing social media as a resource for mental health surveillance' (Conway n.d.)[61]. The analysis below will, however, focus on social media monitoring of content considered relevant for HIV/AIDS risk factors.[62]

Research in this field has as yet received little public attention, possibly due to the fact that it has emerged relatively recently. Moreover, it could be speculated that these research practices were not found to be controversial or problematic by journalists or other observers. In any case, insights have so far mainly been communicated via academic outlets, and targeted at researchers or public health professionals/institutions. Therefore, the arguments brought forward in this context are likewise predominantly established by researchers and not by external observers such as journalists or private individuals. Drawing on Habermas' notion of validity claims, especially with regards to 'normative rightness', but also 'truth' and 'authenticity', the following sections elaborate on the ethical arguments raised in big data-driven approaches to monitoring of HIV/AIDS risk behaviour.

HIV/AIDS risk behaviour refers, for example, to drug consumption which can be hazardous to health, such as the sharing of needles or unprotected sex. To examine how such factors could be monitored via social networking sites, all the projects mentioned above make use of Twitter data. As described in Chapter 3, the microblogging platform broadly allows for open data access. Building on Twitter data, Wang et al. (n.d.) '[…] propose to create a single automated platform that collects social media (Twitter) data; identifies, codes, and labels tweets that suggest HIV risk behaviors'. The platform is meant to be used as tool and service by stakeholders such as HIV researchers, public health workers and policymakers.

The project starts from the hypothesis that certain tweets indicate that individuals intend to or did engage in sex- and drug-related risk behaviour. Some of those tweets can be (roughly) geographically located and enable the monitoring of certain populations (see Young, Rivers, and Lewis 2014). The significance of retrieved data is assessed by combining them with data from established public health surveillance systems as provided by, among others, the US Centres for Disease Control and Prevention (CDC) or the WHO. Wang et al.'s project is particularly focused on automating the processes leading to an identification of potentially relevant data.

In a related paper, the involved scholars acknowledge the importance of preventing their research being linked back to individual persons, since this could lead to stigmatisation (Young, Yu and Wang 2017: 130). For this reason, only a partial list of keywords significant as risk factor indicators has been provided. While stating that '[a] large and growing area of research will be focused on how to address the logistical and ethical issues associated with social data' (130), the authors do not address those issues in detail themselves. However, the project by Young et al. (n.d.; as mentioned before, the scientist was also involved in the study mentioned above) refers explicitly to the relevance of ethical concerns. Methodologically, it moves beyond an exploration of technical challenges. It adds qualitative interviews with '[…] staff at local and regional HIV organization and participants affected by HIV to gain their perspectives on the ethical issues associated with this approach' (Young et al. n.d.).

The two projects highlight typical, insightful approaches to ethical issues in big data research. Concerns regarding the normative rightness and risks of big data-driven studies are framed as challenges to be overcome in future research; they are, however, not seen as reasons to explore beforehand which moral issues may arise. This innovation-driven approach also reflects the conditions under which biomedical and life scientists compete for funding. In the abovementioned cases, it remains to be clarified if and how such research may affect social media users, for example by becoming accused of or associated with presumed HIV/AIDS risk behaviour. But, practically speaking, flagging severe ethical issues may undermine the perceived feasibility and 'fundability' of a research project.

Moreover, an emphasis on ethical questions appears less likely to receive funding in schemes explicitly targeted at software development. At the same time, these dynamics seem related to a lack of ethical guidelines concerning biomedical big data, commonly ensured by institutional/ethical review boards (I/ERB). Ethical decision-making processes for big data-driven public health research operate currently according to negotiated rationales, such as necessity versus the obsolescence of informed consent (see Chapter 3). This also puts involved researchers at risk of public, morally motivated scandalisation and distrust.

Already in traditional *Infectious Disease Ethics* (IRD), a sub-discipline of bioethics concerned with ethical issues regarding infectious diseases, Selgelid et al. (2011) observed comparable tensions between scientists and philosophers, particularly ethicists. While scientists experienced certain moral expectations as unrealistic and oblivious of research realities, philosophers perceived scientists' consideration of ethical issues as naïve. This in turn was countered by scientists with the objection '[…] we are not ethicists, we're just describing an ethical issue we have observed' (Selgelid et al. 2011: 3).

A view of ethics as an 'ex post' perspective is thus not a feature characteristic for big data-driven research, but rather a tendency which can be found in novel, emerging research fields. Moreover, it brings forward the normative claim that ethics cannot be demanded as key, analytic expertise from (data) scientists. Such dynamics have facilitated a 'pacing problem' in innovative research and a '[…] gap between emerging technologies and legal-ethical oversight' (Marchant, Allenby and Herkert 2011). In fast-changing technological cultures, ethical debates often lag behind (see also Wilsdon and Willis 2004). This point hints not only at the importance of strengthened collaboration and mediation between ethicists and scientists, but also at the need for research skills relevant to projects' ethical decision making and increased public outreach.

A recurring ethical, contested issue in this context, as already indicated in Chapter 3, is the question of informed consent. While Young et al. (n.d.) deliberately incorporate stakeholders such as public health professionals and individuals affected by HIV, the role of other users creating data receives little consideration. It has been pointed out that posting content on social media does not necessarily correspond with users' awareness of possible, future uses. Furthermore, users often have little means of privacy management once they opt-in for using certain platforms (Baruh and Popescu 2015; Antheunis, Tates, and Nieboer 2013; boyd and Ellison 2007). Research drawing on such data affects users as it claims access to personal data whose use has not been explicitly authorised by the respective users. This has implications for the societal appreciation of personal autonomy.

The tendency to portray informed consent as neglectable is linked to the common framing of big data approaches as 'unobtrusive', i.e. occurring seemingly without intervening with individuals' activities (see also Zwitter 2014). For example, the scientists involved in the project 'Online media and structural influences on new HIV/STI Cases in the US' (Albarracin et al. n.d.) examined tweets as possible indicators of HIV prevalence in (2079 selected) US counties. Similar to the projects by the PIs Wang and Young, Albarracin et al. also focus on potential links between linguistic expressions on Twitter and HIV prevalence in a population. The authors describe their retrieval of 150 million tweets, posted between June 2009 and March 2010, as '[…] an unobtrusive, naturalistic means of predicting HIV outbreaks and understanding the behavioral and psychological factors that increase communities' risk' (Ireland et al. 2015). In this context, 'unobtrusive' is used in the sense that the data collection does not interfere with users' social media practices.

Implicitly, this interpretation of unobtrusiveness is used as a claim to normative rightness. The normative assumption brought forward in this context is that an approach may be considered unobtrusive because the involved subjects are not necessarily aware that their data are being collected. This claim to the normative rightness and preferability of such approaches is paired with the argument that it produces 'undistorted' and 'better' data, a validity claim to truth. Considering that the latter argument has been challenged as a discursive element of a 'digital positivism' (Mosco 2015) and 'dataism' (van Dijk 2014), these validity claims to normative rightness and truth alike are questionable. Ethically, it implies a misleading understanding of (un-)obtrusiveness which is then presented as advantageous. Methodologically, its claims to reduce distortion appear questionable in the light of research on algorithmic bias (see Chapter 3).

These entanglements between claims to normative rightness and truth are decisive. With regards to *Infectious Disease Ethics*, Selgelid et al. (2011) state that commonly '[r]estrictions of liberty and incursions of privacy and confidentiality may be necessary to promote the public good' (2). But implied measures such as quarantine and mandatory vaccinations usually apply to 'extreme circumstances' (2) or consequences. Moreover, in assessing whether certain ends justify the means, the approaches' effectiveness becomes an important concern. Claims for the normative rightness of social media monitoring for public health surveillance therefore also need to be assessed in light of their claims to effectiveness.

As discussed in Chapters 2 and 3, valid concerns have been raised regarding factors biasing and distorting big data. In the case of the abovementioned studies, two aspects especially should be considered: first, the alterability of corporate big data economies; and second, the fluidity of user behaviour. Both aspects translate into matters of sustainability, reliability, and accuracy. While prominent figures in the field of health informatics such as Taha A. Kass-Hout[63] have declared that "Social media is here to stay and we have to take advantage of it,' […]' (Rowland 2012), neither the platforms nor the corporations owning them are static. Even though Twitter has survived prognoses for its bankruptcy made in 2016 (Giannetto 2015) and it has been said that 'Twitter Inc. can survive' (Niu 2017), the company is struggling to achieve profitability (Volz and Mukherjee 2016).

While one may oppose the possibility that Twitter may be discontinued, given its popularity, it is certainly likely that its data usage conditions will continue to change. This has already occurred in the past, as pointed out by Burgess and Bruns (2012) and Van Dijck (2011). Amendments in Twitter's APIs, making certain data inaccessible, imply that research projects relying on the microblogging platform as their main data source could not proceed as planned. This risk is especially significant when it comes to collaboration with start-ups, as demonstrated by other cases. For example, in February 2016, the Indiana University School of Nursing announced its collaboration with ChaCha, a question and answer online service ('IU School of Nursing and ChaCha partner' 2015).

The platform was available as a website and app. Users could ask questions which were then answered by guides, paid by the company on a contractor basis. It was launched in 2006, received an estimated $43-58 million venture capital within three years (Wouters 2009), first filed bankruptcy in 2013 (ChaChaEnterprises, LLC 2013), and ceased to exist in 2016 (Council 2016). In 2015 the company established a data sharing agreement with the Social Network Health Research Lab (Indiana University, School of Nursing). The researchers received a large (unspecified) dataset of user questions submitted between 2008 and 2012. The aim is/was to analyse questions pertinent to health and wellness, and to explore their implications for public health monitoring. While this one-off data donation still allows researchers to examine the material, follow-up studies involving more recent data would be impossible.

With regards to Twitter and other social networking platforms such as Facebook it has been frequently assumed and argued that privacy is not an ethical issue, because '[…] the data is already public' (Zimmer 2010, 313). In a critical paper on the use of Facebook data for research, Zimmer investigates the unsuccessful anonymisation of a data set and reveals 'the fragility of the presumed privacy of the subjects under study' (314). In a later article, Zimmer and Proferes (2014) oppose the dominant argument that users '[…] have minimal expectations of privacy (Crovitz, 2011), and as a result, deserve little consideration in terms of possible privacy harms (Fitzpatrick, 2012)' (170). When using Twitter, users can choose between either making all their tweets public or restricting access to authorised users. Tweets which are posted publicly are fed into Twitter's partly open data and can be accessed via API. The company itself has access to all tweets, published publicly or privately, as well as metadata, i.e. hashtags, page views, links clicked, geolocation, searches, and links between users (172). Zimmer[64] and Proferes (2014) show that despite Twitter's seemingly straightforward, binary mechanism of public and private tweets, the platform's marketing generally evokes promises of 'ephemeral content sharing'.

As part of the Council for Big Data, Ethics, and Society,[65] established in 2014 as an initiative providing critical social and cultural perspectives on big data, a report by Uršič (2016) shows that in cases where civic users delete tweets or content, this material often remains part of retrieved datasets (5ff.). Coming back to the use of Twitter data for monitoring HIV/AIDS risk factors, the wish to delete personal tweets may occur especially once it transpires how certain content may be interpreted. One should also take into account that not only a platform's appearance, usage conditions and possibilities may be fluid, but that the same goes for users' behaviour. Once aware of the possibility that certain communications (even if only vaguely related to one's sex life, drug consumption, or social drive) may be interpreted as risk behaviour, this could alter users' content production.

Such a development is easily conceivable, given common prejudices towards and the stigmatisation of individuals' suffering from HIV/AIDS. And even without such an explicit intention to adjust behaviour to avoid discrimination, or the impossibility to find an insurer, individuals' interests and practices change. This means that content which might have implied drug- or sex-related risk behaviour may in the foreseeable future take on a different meaning. At this point, it is insightful to remember 'lessons learned' from the discontinuation of Google Flu Trends. In an article on 'big data hubris', Lazer et al. (2014) warn that the constant re-engineering of platforms such as Twitter and Facebook also means that '[…] whether studies conducted even a year ago on data collected from these platforms can be replicated in later or earlier periods is an open question' (1204). In addition, the authors stress the role of so-called 'red team dynamics' resulting from users' attempts to '[…] manipulate the data generating process to meet their own goals, such as economic or political gain. Twitter polling is a clear example of these tactics' (1204).

Comparable dynamics may not only occur due to activities aimed at deliberate manipulation, but also in cases where users react to current events or trends. As early as 2003, Eysenbach (see also 2006) underlined the possibility of 'epidemics of fear'. With this term, the author differentiates between digital data which may reflect that individuals are directly affected by a disease, and those that emerge because users may have heard or read about a health-relevant development. In the case of Google Flu Trends, for example, it is assumed that search queries indicating 'epidemics of fear' have acted as confounding factors, leading repeatedly to overestimations (Lazer et al. 2014, 1204): inter alia during the 2009 H1N1 pandemic (Butler 2013). For the abovementioned projects, aimed at employing social media monitoring as contributing to HIV/AIDS surveillance, this means that models developed based on research need to be constantly evaluated, adjusted, and recalibrated. One reason for this is that linguistic content which has been selected as a signifier of risk-behaviour may subsequently take on different meanings.

This applies to all projects drawing on social media data such as tweets which have been used for monitoring, for example, influenza or cholera,[66] but it seems notably relevant for projects that address stigmatised health conditions. Likewise, the politics behind the selection of certain content which is screened as being indicative for risk behaviour should also be considered. This relates particularly to emphasis on groups that are potentially 'high risk'. If we look for instance at concerns in a different area, regarding 'racial profiling' (Welsh 2007), it has been noted that discriminatory attention towards groups can foster selection and sampling bias. While this is not meant to query that HIV/AIDS research is especially relevant for certain vulnerable groups and individuals, the translation of this knowledge into linguistic criteria for big data-driven research may facilitate sampling biases in the chosen material.

With regards to observational epidemiology, Chiolero (2013) remarks that already in a 'pre-big data era' the trust in large-scale studies occasionally undermined methodological scrutiny. As the author observes, '[…] big size is not enough for credible epidemiology. Obsession with study power and precision may have blurred fundamental validity issues not solved by increasing sample size, for example, measurement error, selection bias, or residual confounding' (Chiolero 2013). Such methodological issues are, however, difficult or even impossible to assess for an external observer, since the ethical concerns regarding stigmatisation led to scientists' decision not to reveal linguistically significant keywords and data.

Similar variations of digital disease detection have also been used in response to natural disasters and humanitarian crises such as the 2010 Haiti earthquake and the subsequent cholera outbreak (Meier 2015; Chunara et al. 2012). On Twitter's tenth anniversary, UN Global Pulse praised the platform as '[…] one of the central data sources here at Global Pulse during our first years of implementing big data for development programs' (Clausen 2016). But Twitter is only one of many platforms which the initiative aims to involve in its vision of data philanthropic, public-private collaborations for development (Kirkpatrick 2016). Humanitarian initiatives such as the Ushaidi Haiti Project (UHP) also gained significant insights into which and where medical support and aid was needed in the aftermath of the 2010 Haiti earthquake. It did so by analysing a variety of (non-)digital sources. UHP established a digital map, bringing together: geographically located tweets; SMS sent to an emergency number; emails; radio and television news; phone conversations; Facebook posts and messages; email list-contributions; live streams and individual observation reports (Meier 2015, 2ff.).

Privacy concerns regarding data retrieved from Twitter, as indicated above, are commonly seen as unreasonable. Still, there are researchers who have stressed users' expectation of privacy even under these conditions (Zimmer and Proferes 2014). But how do we know how users perceive and are affected by research using their data, given that informed consent is neglected and other qualitative data on the issue are still largely missing? This issue becomes even more complicated when looking at social networking sites and content for which the differentiation between public and private is more ambiguous, as in the case of Facebook 'likes' and other digital interaction data.

Unhealthy Likes: Data Retrieval Through Advertising Relations

Social media data are not always as accessible as in the case of Twitter. In some cases, big data access is granted exclusively or under more restrictive conditions. Researchers who intend to use such data need to acquire access in ways defined by the respective platforms and the corporations that own them. This has been achieved by establishing private-public partnerships, that is: collaboration between employees (potentially researchers) of tech corporations and academics working at universities or public health institutions.

For example, platforms such as Google Flu Trends have been based on collaboration between scientists from the United States CDC and Google employees (Ginsberg et al. 2009)[67]. Similar research using Yahoo search queries as data for influenza surveillance involved a Yahoo Research employee (Polgreen et al. 2008). The first mentioned author of the 'emotional contagion experiment' (Kramer, Guillory, and Hancock 2014; see *Informed Consent* in chapter 3 of this book) works for Facebook's Core Data Science Team. It has been discussed already that the conditions for establishing such partnerships are largely opaque. They depend on corporate preferences and individual negotiations, often in favour of well-known and networked elite universities.

As an alternative to such collaboration and institutional dependencies, researchers have explored a form of data access which allows for possibilities comparable to the described Twitter data: they place themselves in the position of advertising customers. This does not necessarily mean that they pay for retrieved data, even though this has also been the case. Either way, researchers do collect such data via channels originally designated for advertising and marketing purposes. One of the earliest examples of this is an approach which Eysenbach called the 'Google ad sentinel method'. The epidemiologist was able to demonstrate '[…] an excellent correlation between the number of clicks on a keyword-triggered link in Google with epidemiological data from the flu season 2004/2005 in Canada' (Eysenbach 2006, 244). But obviously such data were and are not openly accessible.[68]

Eysenbach described his approach as a 'trick' (245), since the actual Google search queries were not available to him. Instead, he created a 'Google Adsense' commercial campaign, which allowed him to obtain insights into potentially health indicative data. His method was not able to obtain actual search query quantifications, but he was able to factor in those users who subsequently clicked on a presented link. When (Canadian) Google users entered 'flu' or 'flu symptoms', they were presented with an ad 'Do you have the flu?', placed by

Eysenbach. The link led to a health information website regarding influenza. As an alleged advertising customer, Google provided the researcher with quantitative information and geographic data for users who clicked on the placed ad. When relating these clicks to data from the governmental 'FluWatch Reports' (provided by the Public Health Agency Canada), he detected a positive correlation between the increase of certain search queries and influenza activities. Eysenbach describes his approach as a reaction to a 'methodological problem [which] lies in the difficulties to obtain unbiased search data' (2006, 245). The ethical implications of this method and of the conditions leading up to its development are up for debate, however.

The use of data meant for advertising customers has been comparatively less common, and was predominantly applied to North American users. Research involving Facebook's social data is noteworthy. Advertising on Facebook has been used for recruiting study participants (Kapp, Peters, and Oliver 2013). In such cases, researchers had to pay for the placed ads and received, in addition to responses from interested individuals, access to the data generated in this process.[69] However, scientists have also registered as business customers for Facebook's advertising and marketing services – which disclose some data freely, without any necessary payment. Based on the latter approach, Chunara et al. (2013) and Gittelman et al. (2015) explored how Facebook's developer platform, available APIs and data may be utilised as means of public health surveillance.

In terms of relevant actors, it makes sense to first look at the specific stakeholders involved in both papers. The paper by Chunara et al. (2013) is based on collaboration between academics working at US universities. The team consulted an (unspecified) advertising company for information on Facebook's data retrieval possibilities and conditions.[70] Gittelman and his co-author Lange were/are (in 2017) both employed at Mktg, Inc. which presents itself as 'lifestyle marketing agency'. Gittelman is the company's 'president CEO'.[71] Further co-authors are employed at the CDC (National Center for Chronic Disease and Health Promotion) and USDA National Agricultural Statistics Service. These constellations are an insightful indication of the expertise needed and merged in such research.

Expertise in big data analytics has been extensively cultivated in marketing and advertising contexts. Related actors possess skills which are crucial for employing social media data. This has enabled them to participate in research involving big data, complementing the expertise of researchers specialised in, for example, public health. In these contexts – involving public-private collaboration or consultancy relations – marketing expertise becomes an asset in public health research. On the side of the users, it also means that Facebook content posted, exchanged or clicked on for entertainment purposes and social interaction is turned into health relevant information. In this case, Facebook users whose data were retrieved for relevant studies are particularly crucial stakeholders. In both abovementioned cases, as noted earlier, these are users located in the US.

Chunara et al. (2013) assess how various Facebook data may contribute to public health surveillance of obesity. According to the authors, the availability of geographically specific data makes the social network a particularly valuable source. Facebook allows potential advertising customers to pre-assess and choose potential target groups '[…] based on traits such as age, gender, relationship status, education, workplace, job titles and more.'[72] This specifically includes information on geographical location, interests (e.g. hobbies or favourite entertainment) and behaviours (e.g. purchase behaviours or device usage). Through Facebook for Developers and its advertisement/marketing platform, such data were accessed by Chunara et al. (2013). As the authors describe:

> 'The platform provides the number (found to be updated approximately weekly) of users who fall under the selected categories and demographics at the resolution of zip code, city, state, or country including surroundings at varying geographic radii. Categories are determined through individuals' wall postings, likes and interests that they share with their Facebook friends and through which they create a social milieu.' (Chunara et al. 2012, 2)

Categories can be accessed as aggregated user profiles, based on certain areas of indicated interests and habits. Chunara et al. selected particularly the categories 'health and wellness' and 'outdoor fitness activities' as relevant indicators to assess obesity prevalence. Social media data focused on these categories was then related to data from the CDC's Behavioral Risk Factor Surveillance System. The authors found '[…] that activity-related online interests in the USA could be predictive of population obesity and/or overweight prevalence' (Chunara et al. 2013, 6). While the authors do not present this as a surprising outcome as such, they frame their study as a contribution to identifying viable, novel methods and complements in public health surveillance. Potential limitations are discussed carefully (ibid, 4-6); however, these are depicted as methodological challenges rather than reasons for ethical concerns.

The abovementioned study involves diverse social data sources, for example content such as wall postings, likes and indicated interests. In comparison, Gittelman et al. (2015) focus on 'likes', i.e. users' clicks on Facebook's famous like-button. This button is predominantly read as an expression of interest in as well as support and sympathy for certain content. The authors examine how the data emerging from users' 'liking' of content may act as potential health indicators for mortality and disease rates, as well as so-called lifestyle behaviour. Comparable with the approach of Chunara et al. (2015), they use aggregated data of users, sorted by zip code. These users 'liked' certain items, falling under certain categories.

The data are retrieved through Facebook's marketing/advertisement platform for developers. As the authors explain, they selected three main categories from the available eight overarching categories – events, family status, job status, activities, mobile device owners, interests, Hispanic[73], and retail and shopping – relevant to US audiences. They chose 'activities' and 'interests' because these include the sub-categories 'outdoor fitness and activities' and 'health and well-being', as assumed factors for self-care and physical activity. The category 'retail and shopping' was selected as an indicator of socio-economic status (SES), which is linked to health- conscious behaviour and financial opportunities to realise a healthy lifestyle (Gittelman et al. 2015, 3).

The data obtained from Facebook were then correlated with public health data from the US National Vital Statistics System (e.g. on mortality rates, disease prevalence, and lifestyle factors) and the US Census, as well as self-reported data from the Behavioral Risk Factor Surveillance System (BRFSS). The latter includes information on habits such as smoking and exercise, the health insurance status ('insured'), and health conditions such as diabetes, prior heart attack or stroke. Based on correlations between these sources and social media data, the authors argue that, in combination, Facebook likes and socio-economic status (SES) indicators, for example income, employment, education information, can predict the tested disease outcomes (see Gittelman 2015, 4). Moreover, they stress the behavioural significance of such data by portraying 'likes' 'as a measure of behaviour' and determining 'the behaviors that drive health outcomes.' (ibid).

In this sense, Facebook data are not merely presented as indicators of existing health conditions, but also of likely, future behaviour. The latter assumption, in terms of technological promises, reduces the complexity of health-relevant behaviour to schematic categories which have been conceptualised for advertising purposes.[74] Moreover, it does not take into account the fluidity of social media as such – which has been demonstrated, for instance, by Facebook's 2016 introduction of 'like' alternatives called 'reactions'.[75] As opposed to emphasised, ambitious promises, an ethics section and reflections on eventual moral concerns are entirely missing from Gittelman et al.'s (2015) article.

For the US, access to health relevant information via social networking sites such as Facebook is possible due to the lack of legal frameworks protecting users' rights to certain big health data. With regards to medical privacy, the Electronic Frontier Foundation (EFF) stresses that social networking sites and other online services pose severe risks and threats to individuals' control of personal data. This applies particularly to users located in the US. The EFF details this situation and its implications as follows:

> The United States has no universal information privacy law that's comparable, for instance, to the EU Data Protection Directive. [...] The baseline law for health information is the Health Insurance Portability

and Accountability Act (HIPAA). HIPAA offers some rights to patients, but it is severely limited because it only applies to an entity if it is what the law considers to be either a 'covered entity' – namely: a health care provider, health plan, or health care clearinghouse – or a relevant business associate (BA). This means HIPAA doesn't apply to many entities who may receive medical information, such as an app on your cell phone or a genetic testing service like 23andMe (Electronic Frontier Foundation n.d.).

This also implies that US users' Facebook or Twitter data, despite their actual use as health indicators, are so far not protected under HIPAA.[76] In Europe, the data protection directive mentioned in the above quote has been meanwhile replaced by the EU General Data Protection Regulation (GDPR). A directive sets out objectives to be achieved by all EU countries; in contrast, a regulation is a legally binding legislative act. The GDPR was adopted in April 2016 and will be fully implemented by the end of May 2018 (see also Morrissey 2017). Its consistent application across the EU will be overseen by the European Data Protection Board (EDPB). The GDPR has been described as an important step towards safeguarding European users' rights and privacy in a global data economy. At the same time, businesses have been concerned about compliance requirements and practical challenges, implying economic disadvantages. Moreover, in response to earlier/draft versions of the GDPR, biomedical researchers, notably epidemiologists, raised the issue that parts of the regulation allow for interpretational leeway and could lead to overly restrictive informed consent requirements (Nyrén, Stenbeck and Grönberg 2014, 228ff.).

As so often, data protection turns out to be negotiated as a trade-off between public wellbeing and broader benefits, a society's capacity for innovation, and individual rights. With regards to Europe, tensions between users' rights and data as a driver for innovation have been extensively considered in documents released by the European Data Protection Supervisor (EDPS), an independent EU institution. It has been pointed out, from an innovation and research perspective, that the legal restrictions implemented in this field may impede the productivity of research and innovation.[77] Even in the EC General Data Protection Regulation (GDPR) exceptional status is granted to the use of personal data in certain situations, referring to the need of weighing the public good and individual rights:

> 'Such a derogation may be made for health purposes, including public health and the management of health-care services, especially in order to ensure the quality and cost-effectiveness of the procedures used for settling claims for benefits and services in the health insurance system, or for archiving purposes in the public interest, scientific or historical research purposes or statistical purposes.' (The European Parliament and the Council of the European Union 2016, 10; §52)

Such retrenchments are necessary; they open up possibilities for highly relevant and needed health research. But they also require further ethical considerations on the question of which cases derogations are reasonable. It is therefore problematic if ethical reflections on these issues are neglected when it comes to big data-driven health surveillance. Such research fails to address, in terms of normative rightness, why informed consent appears dispensable under certain conditions. This also means that public debate on this issue is a priori uninformed, a factor that is indispensable for the formation of valid social norms, according to Habermas.

One of the normative arguments recurring in various big data-driven health studies and justifications is the emphasis on the 'cost-effectiveness of the procedures' (see also the quote above). Already in his early 2006 study, Eysenbach stressed the timeliness and accuracy of what he called the 'Google Ad Sentinel method'. He also pointed out its cost effectiveness compared to more traditional approaches to influenza surveillance (Eysenbach 2006, 244; see also 246). Similar statements can be found in the papers by Gittelman et al. (2015) and Chunara et al. (2013). Gittelman et al. describe their method as a contribution that 'directly affects government spending and public policy' and comes at 'a fraction of the cost of traditional research' (2015, 7). Chunara et al. (2013) stress that their big data-driven research offers 'a real-time, ease of access, low-cost population-based approach to public health surveillance' (2013, 6).[78] This emphasis on financial benefits needs to be seen in the context of health care systems which are under ever increasing pressure to economise and reduce costs (Kaplan and Porter 2011). The authors strengthen the (misleading) assumption that big data provide a solution for this issue. This conclusion needs to be urgently mitigated by re-emphasising the societal costs looming due to an inordinate, naïve reliance on technological promises, promoted by internet and tech corporations. These costs are related to public health monitoring platforms modelled on fluid big data economies (see the previous sub-chapter on Twitter data); a conceptualisation of users' as static, non-reflective entities; and a negligence of algorithmic biases and recalibration needs.[79]

Public Health and Data Mashups

The studies mentioned and described above, involving Facebook, Twitter, and Google data, have in common that they initially focus on stages of methodological exploration. The authors examine how available data could be analysed and used for public health surveillance. Ultimately though, in most cases, such investigations strive for technological utilisations of their methodological insights. Most of them have a concrete development aspect.

This is obvious in the case of projects funded as part of the US BD2K grant scheme 'Targeted Software Development', which applies to Wang et al. (n.d.) and Young et al. (n.d.). As part of the interconnected projects, the two PIs are involved in creating a platform which automatically retrieves, analyses, and visualises Twitter data indicative of HIV/AIDS high-risk behaviour (Wang et al. n.d.). Also, the UN Global Pulse Labs are developing practical applications such as the 'Haze Gazer', a crisis analysis tool supported by the government of Indonesia,[80] and implementing public dashboards for, among other uses, 'Monitoring in real time the implementation of HIV mother-to-child prevention programme' in Uganda.[81] Chunara and Brownstein, both of whom contributed to the aforementioned paper on monitoring obesity prevalence through Facebook data (Chunara et al. 2013), are part of a team engaged in various explorations and practical applications of 'digital disease detection' (Brownstein, Freifeld and Madoff 2009). In interdisciplinary collaboration with biomedical and computer scientists, they notably developed a platform called HealthMap. This has been described by *Wired* as a manifestation of Larry Brilliant's wish and vision for a freely accessible, online, and real-time public health surveillance service (Madrigal 2008; see also Chapter 4).

HealthMap is an example of data mashups, which are increasingly common. These are websites which select and combine data from diverse online sources (Crampton 2010, 25ff.). In the case of public health surveillance services, they are often combined with geographic maps. Cartographic visualisations facilitate epidemiological insights into the spatial patterns and spreading of infectious diseases. Maps may support public health professionals in assessing how quickly a disease spreads and which spatial patterns emerge. At the same time, they serve as accessible tools for communicating disease information to the public. Spatial analyses and visualisations of epidemics are part and parcel of public health surveillance (see also Ostfeld et al. 2005).

Already in the mid-1990s, Clarke et al. examined the potential use of emerging Geographic Information Systems (GIS) (i.e. locative processing and visualisation tools) in epidemiology. The authors stressed the promises coming along with such developments: 'GIS applications show the power and potential of such systems for addressing important health issues at the international, national, and local levels. Much of that power stems from the systems' spatial analysis capabilities, which allow users to examine and display health data in new and highly effective ways.' (Clarke et al. 1996, 85) The use of data map mashups is a continuation of previous public health surveillance practices, but opens up novel possibilities and challenges.

The use of data map mashups for public health surveillance has been explored since the mid-2000s. The public services EpiSPIDER (Tolentino et al. 2007; Keller et al. 2009) and BioCaster (Collier et al. 2008) mapped data retrieved from various online sources, such as the European Media Monitor Alerts, Twitter, reports from the US CDC and the WHO. The selected information was then presented in Google Maps mashups. Google Flu Trends (GFT, Ginsberg et al. 2009) can be considered Google's in-house solution for Brilliant's vision of an online disease surveillance system. (Brilliant was involved in the project and paper himself). While GFT aimed at predicting influenza intensities based on search queries that were previously correlated with traditional health surveillance data, HealthMap's objectives are more diversified in terms both of the diseases included and the data. In both cases though, the retrieved and selected data are/were[82] presented in an interface integrating Google Maps.[83]

For the creation of HealthMap, epidemiological expertise, data science, and bioinformatics had to go hand in hand. In terms of directly involved stakeholders, the platform was developed by interdisciplinary teams of epidemiologists, computer scientists (particularly bioinformaticians), and data scientists. It was launched in 2006, enabled by research from an interdisciplinary team at Boston Children's Hospital, with epidemiologist Brownstein and computer scientists and biomedical engineer Freifeld in leading roles. The project has been extensively documented by involved scientists in publications in leading academic journals (see e.g. Brownstein et al. 2008; Freifeld et al. 2008; Brownstein and Freifeld 2007).

HealthMap received funding from multiple corporations, for example a grant of $450,000 by Google's 'Predict and Prevent' initiative as well as from Unilever, Amazon, and Twitter, and foundations such as the Bill and Melinda Gates Foundation and the Skoll Global Threads Fund. It was also provided with financial support from governmental agencies such as the US Defense Threat Reduction Agency (DTRA), the CDC, the NIH National Library of Medicine, and the Canadian Institutes of Health Research (CIHR). Visually, the interface is dominated by Google Maps: in this map, health relevant information – such as news items on disease outbreaks or tweets concerning disease developments in a certain region – are located. The selection process is automatised, in that certain sources are monitored by default and it is algorithmically determined which content will be included. Depending on the website users' location, a smaller text-box on the right indicates potential 'Outbreaks in current location' which are clustered into twelve disease categories. [84]

HealthMap combines data which are retrieved by scanning multiple sources. Among them are the commercial news feed aggregators Google News, Moreover (by VeriSign), Baidu News and SOSO Info (the last two are Chinese language news services), but also institutional reports from the World Health Organisation and the World Organisation for Animal Health, as well as Twitter.[85] The platform utilises global sources and is not limited to a particular country. These are authored by public health institutions or news outlets/journalists. Before being published, such sources are commonly subject to selection and verification processes during which their quality and correctness is assessed. This applies particularly to organisations such as the WHO, but is also the case for quality journalism outlets (Shapiro et al. 2013). In contrast, microblogging platforms such as Twitter also contain information from individual users. Although this latter source of information may be more current, it is also more difficult to verify (Hermida, 2012). Apart from automatically retrieved social media content, users can also send individual reports: this can either be done through the website's 'Add alerts' function (which is part of the top menu), by email, text message, phone call (hotline), or by using the mobile app *Outbreaks Near Me*.

News items are a particularly dominant type of data, mostly retrieved from news aggregators, with Google News items being especially prevalent. Therefore, being included in such aggregators enhances the chance for (health-indicative) news items to be presented in HealthMap. These aggregators, maintained by global tech corporations, play an important role as gatekeepers, defining in- and exclusion. In this sense, research concerning the gatekeeping function of such aggregators is highly relevant to projects such as HealthMap, and may be used to assess the implications of such an approach (Weaver and Bimber et al., 2008).

While drawing on news aggregators seems to be a technically feasible/preferable solution, this approach raises questions regarding the selection criteria relevant to utilised big data sources. The presented data go through multiple forms of automated selection: first, they are defined by, for example, the Google algorithm that determines more generally which sources are included in its *News* service. Second, they are subject to an automated process in which the HealthMap algorithm selects information which is considered relevant for disease detection.

In combination with the funding the project received, the used content poses questions regarding eventual conflicts of interests and emerging dependencies. Exaggerating somewhat, technology editor Reilly (2008) remarked of HealthMap: 'We can't officially call the program Google Disease(tm). But that's essentially what HealthMap is.'[86] In an interview, Google 'Predict and Prevent' director Mark Smolinski commented on HealthMap and the decision to provide funding: 'We really like their approach in that they are trying … a really open platform,' […] 'Anybody can go in and see what kind of health threats are showing up around the world' (Madrigal 2008).

The fact that Google material is being used provides the corporation with positive public exposure. It links the company's (branded) content to techno-scientific innovation as well as the well-established perception that public health surveillance is an important contribution to societal wellbeing. Whether the use of Google data is, methodologically speaking, the ideal approach for HealthMap remains to be explored. Ethically, the emerging dependencies may result in stakeholder constellations between data providers and scientists which affect future decision making. This latter effect has already been described with regards to pre-big data industry funding (Lundh et al. 2017; Bekelman, Li and Gross 2003).

The dominance of Google News items in large parts of Europe and the US is also likely related to a main methodological challenge already addressed by the scientists involved in the creation of HealthMap. With regards to the used 'web-accessible information sources such as discussion forums, mailing lists, government Web sites, and news outlets', Brownstein et al. (2008) state that '[w]hile these sources are potentially useful, information overload and difficulties in distinguishing 'signal from noise' pose substantial barriers to fully utilizing this information'. This concern refers to the challenge of selecting relevant data, but it should also be seen in the context of different data sources providing varying amounts of data.

Considering Google News' extensive, ongoing data collection and capacities, sources which provide quantitatively less input run the risk of being overlooked – in this case not by the algorithm, but by those users trying to make sense of visualised data. What is happening here can be (structurally) compared with the common experience of a Twitter user who starts following very vocal corporate, political or governmental account, for example. The constant 'noise' of such quantitatively dominating actors is likely to impede one's perception of other, relevant information sources.

Dependencies and potential conflicts of interest concern the content which is mapped, but also the Google map itself. The fact that content is placed in Google Maps also raises issues concerning sustainability, similar to those dynamics described for Twitter data. Critical geographers were also among the first to tackle the sensitivity of big data and locative information (Dalton and Thatcher 2014; see also Chapter 1). They have cautioned against uncertainties when relying on corporate services in *neogeography*. The latter notion implies that maps are created and processed by actors who are not trained cartographers, but participate in map-making with the help of cartographic online services (see also Rana and Joliveau 2009, 79).

There are various mapping services, such as Google (My) Maps, the above-mentioned Ushaidi platform, or the free and open source project Open Street Map, which enable non-cartographers to map information or even to create cartographic surfaces. It has been highlighted, though, that these participatory mapping approaches are still subject to regulations defined by the map hosts. This is especially relevant in cases where the cartographic material is owned by corporations such as Google. Various authors have challenged optimistic assumptions of a 'participatory mapping culture' and its democratisation. They point out that neogeographic practices are defined by access to the internet and digital content as well as digital skills and literacy.

Haklay (2013) criticises the starry-eyed promise that neogeography 'is for anyone, anywhere, and anytime'; instead, the author argues that looking at the actual practices exposes sharp divides between a technological elite and 'labouring participants' (Haklay 2013, 55).[87] In addition to such issues of accessibility and expertise, there are new forms of dependency which are related to the dominance of global media corporations: 'One of the more curious aspects of Neogeography is the high dependency of much activity on the unknown business plans of certain commercial bodies providing API's for mapping.' (Rana and Joliveau 2009, 80) This also has an influence on the sustainability of projects relying on commercial APIs, since the conditions for using them may change – as also remarked with regards to prior research approaches.

Potential conflicts of interests and dependencies in big data-driven health projects should be placed in the context of broader ethical considerations for datafied societies. Calling attention to seminal changes emerging in research connected to global tech corporations, Sharon (2016) argues that since unfolding '[…] power asymmetries may affect the shaping of future research agendas, they deserve greater critical attention from medical researchers, ethicists and policy makers than is currently the case.' (564).[88] It is striking that such concerns are rarely an integral part of techno-scientific explorations of big data-driven public health research. What can be considered 'disruptive communicative action' in Habermasian terms does occur, for instance in those critical contributions which I have continuously referenced above. But these disruptions are never moved toward a level of 'higher' argumentative discourse.

An engagement with ethical issues that takes the side of those involved in big data-driven public health surveillance is reduced to justifications of research practices, or in some cases is even missing. In those, still exceptional cases, where such validity claims to normative rightness are raised and challenged, a discursive divide between those arguing from an ethical and those from an innovation-driven, methodological perspective prevails. Ethical arguments appear to unfold in distinct spheres rather than in actual dialogue. From a discourse ethics perspective, this is problematic, since it weakens the validity of social norms and moralities crucial to respective research approaches. In the following, final Chapter 6, I will elaborate on this conclusion by tying it back to the critical perspectives and theory introduced in Chapter 2.

Emerging (Inter-)Dependencies and their Implications

In big data-driven health research, entanglements between academic studies and market-dominating tech/internet corporations have emerged. This is in part related to the tendency that access to online data is increasingly controlled by these companies. Research projects drawing on, for example, social media data depend on collection and access conditions defined by internet and tech corporations. This is also linked, however, to tech corporations' philanthro-capitalist engagement in funding and encouraging research at the intersection of public health and tech-driven innovation. Tech-related topics, development and data science approaches in health research are supported through corporate data, analytics and grant schemes.

How data are retrieved by internet/tech corporations reflects certain norms and values. Big data-driven health research that uses data collected under corporate conditions, runs the risk of echoing and normalising these values and norms as they become decisive conditions for projects' data retrieval. In consequence, this research also reinforces the moral credibility of corporate approaches to users' data by showcasing big data's contribution to societal well-being and public health.

These tendencies have crucial implications for research ethics and integrity. It is particularly notable that studies involving big data tend to diminish possibilities through which affected actors could voice their (dis-)approval. Relevant stakeholders, in particular data subjects, are barely involved in negotiations of norms relevant to data retrieval or use. Informed consent is abandoned, mostly without questioning the appropriateness to do so for specific studies.

From a discourse ethics perspective, the validity of moral norms in big data-driven health research is assessed by asking how they were created in formative discourse (see e.g. Habermas 2001 [1993], 1990). Habermas proposes that the validity of norms depends on whether their assertion safeguards the autonomy

---

How to cite this book chapter:
Richterich, A. 2018. *The Big Data Agenda: Data Ethics and Critical Data Studies.*
Pp. 91–103. London: University of Westminster Press.
DOI: https://doi.org/10.16997/book14.f. License: CC-BY-NC-ND 4.0

of all affected individuals. As a 'counterfactual idealization' (Rehg 2015, 30), his theory is meant to guide and assess (moral) reasoning. 'Justice' is seen as a key dimension of validity for moral discourses; valid norms are those ensuring justice. Habermas' theory has been frequently criticised as utopian. But even though its main normative principles may be ultimately out of reach, they provide reference points towards which (moral) reasoning may orient itself.

Addressing the validity of those social norms guiding big data-driven health research is highly relevant, as ethico-methodological changes in this field compromise many long-established research principles, such as informed consent. As described in Chapter 2, my analysis addresses two main issues concerning big data-driven health research, derived from critical data studies, pragmatist ethics and Habermasian theory: what are the broader discursive conditions, including key stakeholders and factors shaping their views? Which ethical arguments and validity claims have been brought forward? In this chapter, I reflect on the implications of observations and arguments presented in response to these questions in Chapters 4 and 5: stakeholders, discursive conditions and validity claims.

Stakeholders, Discursive Conditions, Validity Claims

Stakeholders

With regards to affected actors, I maintain that there is currently an imbalance and lack of formative discourse defining the ethics and social norms of big data-driven health research. Emerging data practices and ethics are criticised by academics and (occasionally) data activist groups, such as the Electronic Frontier Foundation. But often these debates are carried out in response to big data-driven approaches, rather than being foregrounded by involved researchers themselves. Moreover, there is little formative dialogue between researchers exploring novel approaches and those challenging ethical assumptions made with this research. There is also little discursive involvement when it comes to affected, civic individuals whose data are (or could be) used (Lupton 2016; Metcalf and Crawford 2016). This issue stresses the relevance of enhanced efforts in communicating relevant scientific developments and ethical dimensions of big data-driven research in public health domains.

Such efforts are crucial for fostering individuals' possibilities to voice concern or approval. There is an urgent need to facilitate civic insights and possibilities for formative moral discourse regarding emerging, big data-driven research approaches. This observation also corresponds with what Kennedy and Moss (2015) conceptualise as a much-needed transition towards approaching data subjects as 'knowing' rather than merely 'known publics'. The authors criticise current data practices for addressing publics mainly as passive data subjects, as they are primarily aimed at making sense of datafied individuals (see also Zwitter 2014). Instead data should be used to '[…] help members of the public to understand public issues and each other better, such that more informed and knowing publics may take shape' (Kennedy and Moss 2015, 8). In the case of big data-driven health research such an understanding can only (potentially) occur if research methods are made available for debate in accessible and apprehensible ways. Such research might then also call attention to how personal and sensitive users' digital data really are.

Following Habermas' principles of discourse and universalisation, the only possibility to justify or counter norms which are decisive for big data-driven health research – for example the negligence of informed consent – is to ensure individuals' engagement in practical discourse. Without enhanced investments in involving affected individuals discursively, emerging possibilities for big data access amplify alienation between researchers using, and individuals contributing, data. In many cases, this implies a lessened involvement of affected individuals in relevant discourses and a weakened validity of the moral norms at the heart of such academic research. This is particularly noticeable when looking at debates concerning the role of informed consent.

Scholars involved in and observing big data research have controversially discussed whether the negligence of informed consent is indeed morally reasonable or merely technologically induced in big data-driven research. Informed consent is dismissed by those engaged in big data-driven research as superfluous for studying data subjects, as a relic of obsolete data retrieval conditions and as a now avoidable source of bias. For those defending informed consent, however, informed consent is an indispensable tool for safeguarding the autonomy and dignity of affected individuals. Undoubtedly, informed consent does not perfectly match Habermas' idealised principles and idea of formative discourse. Yet it functions as a research element aimed at approximating conditions for collectively formed, valid and just norms which are ethically decisive for scholarly practices.

By relinquishing informed consent, scholars remove means for involving individuals in a discourse of normative approval or disapproval. In this sense, studies using big data and eschewing informed consent lack forms of discursive involvement fostered in earlier research approaches. In Habermasian terms, such studies move further away from conditions facilitating valid norms '[…] that meet (or could meet) with the approval of all affected in their capacity as participants in a practical discourse' (1990, 66). Current big data-driven research approaches tend to cut out informed consent as an established form of discursive engagement of affected individuals. They also commonly fail to implement alternative possibilities for discursive negotiations of this moral norm.

One of the still rare cases in which such an attempt has been made is the study by Young et al. (n.d.). As described in Chapter 5, their project aims at creating a platform for monitoring tweets which may indicate health related high-risk behaviour in a population. At the same time though, they conduct interviews with individuals working with HIV organizations, as well as participants affected by HIV, on ethical issues regarding the taken approach. As indicated above, whether such approaches are indeed an acceptable alternative to informed consent has been questioned. Nevertheless, such strategies indicate how alternative means for shaping the discursive conditions for public opinion formation and the involvement of affected individuals can be explored.

Discursive conditions

With regards to discursive conditions, I argue that by engaging in big data-driven health research without foregrounding potential risks and ethical issues, scholars facilitate discouragement of discursive, civic involvement. By failing to stress their awareness of potential controversies, they moreover risk scandalisation and increased public mistrust towards emerging, data-driven research approaches. Researchers present the use of big data from a societal position to which the highest moral standards are supposed to apply. They rely heavily on their perception as acting in the interest of the public (Van Dijck 2014). Public trust has been acknowledged as crucial to scientific research practices and moral values in democratic societies (Wynne 2006; Kelch 2002). When using certain kinds of big data in academic research, scholars assert the moral adequacy of norms relevant to their research. At the same time, they assert the appropriateness and value of (corporate) practices needed to acquire the used data.

Mobilising and drawing on the public trust which is widely placed in academic research,[89] they likewise suggest that public scrutiny of big data practices is not necessary. In doing so, however, they fail to facilitate a better public understanding of how personal and sensitive social media data may be. This both fosters the abovementioned negligence of stakeholders and in turn, weakens the validity of morals crucial to research. When ethical debates happen, they often have an effect on public trust in science. The importance of ethical foresight has therefore also been stressed with regards to avoiding a 'whiplash effect', i.e. (over-)regulations due to extremely negative perceptions of scientific and technological developments (Mittelstadt and Floridi 2016, 305ff.).[90]

These risks are related to competitive funding systems for public health research in which not only governmental grant schemes, but internet and tech corporations have come to play a distinct role. I elaborated in Chapter 4 that internet/tech corporations engage in supporting and funding projects investigating how digital technologies and big data may be employed. They particularly target domains considered as beneficial and relevant to societal development, notably public health research. This also means that such companies play a role in shaping contemporary research agendas. These corporate funding opportunities incentivise studies exploring how technological developments more generally, and big data specifically, can be used in research. Furthermore, such funding schemes, and especially research taking place within corporations, are not overseen under the same conditions as research funded through governmental grant schemes (concerning, for example, ethical review).

Significant interest in the intersection of technology and big data, science and public health does not only apply to corporate funding and support. Governmental, (inter-)national funding schemes reinforce investments in tech and big data-driven research. The need to acquire funding to conduct research is a common prerequisite for contemporary scholarship (Hicks 2012; Benner and Sandström 2000). The conditions, criteria and ramifications of governmental funding schemes have been widely criticised, though (Geuna 2001). Berezin (1998) even famously stated that '[a] random lottery among the competent applicants would do equally well and, perhaps, even better, because it at least avoids the bias of sticking to current fads and fashions so typical of the conventional APR of research proposals' (10). Moreover, the significance of lobbying and policy developments for research trends has been pointed out (Parsons 2004).

Yet while also being far from complying with the Habermasian ideal of discursive conditions taking into account all potentially affected individuals, in democratic societies, governmental funding schemes aim at reflecting democratic values and decision-making processes. In contrast, corporate funding instruments are part of the rise of philanthrocapitalism, and of what Horvath and Powell (2016) termed 'disruptive philanthropy' (89; see also Alba 2016). It is characteristic for internet and tech corporations engaged in philanthrocapitalist strategies to invest in projects promising to improve societal wellbeing through technological innovation.

Corporate interests and agendas, such as technology and its benefits, are merged with domains that are associated with widely accepted moral values, notably related to public health. In most of these cases, the charitably invested money will not be taxed in ways which would have led – at least partly – to its contributing to governmental programmes guided by democratic values (Alba 2016; Horvath and Powell 2016). When research funding is linked to corporate interests, efforts aimed at democratic decision-making processes concerning research grants and schemes are undermined. Not only interdependencies, but also dependencies and conflicts of interest emerge: corporations are providing data, analytics, interfaces and grants for studies that are relevant to their economic interests and public image.[91] These dynamics raise the question to what extent tech corporate agendas are getting 'baked into' research projects.

Complex interdependencies emerge especially around those projects using data and tools from the tech corporations that fund them. Sharon (2016a) reminds us that '[…] insofar as the devices and services that generate, store, and in some cases analyze these data are owned by commercial entities that are outside traditional health care and research, we also should be attentive to new power asymmetries that may emerge in this space, and their implications for the shaping of future research agendas'. These constellations result in dependencies and potential conflicts of interest which may be difficult for involved scientists to resolve. The issue also relates back to the abovementioned concerns that the merging of corporate data retrieval and academic research may be hazardous to the reputation of the latter.

Public-private partnerships, for example between university projects and tech corporations, affect the public perception of both. Corporations providing data or grants benefit from associating themselves with the relevance and contributions of scientific endeavours. At the same time, scientists may be increasingly associated with moral concerns pertinent to corporate practices. With regards to initiatives using big data, the UK Science and Technology Committee (2015) stresses that misuses and leaks of data have fostered public distrust towards governmental as well as corporate practices: referring to studies conducted by pressure groups such as Big Brother Watch Ltd., the report notes '[…] that 79% of adults in the UK were 'concerned' about their privacy online, and 46% believed that they were 'being harmed by the collection of their data by large companies' (Science and Technology Science and Technology Committee, House of Commons 2015).

These assessments partly contrast with a 2014 Eurobarometer survey on 'Public perception of science research and innovation' and the European Commission's report published on its results. In response to this report, Floridi (2014) summarises its main results and suggests possible interpretations:

> 'As a priority, data protection ranks as low as quality of housing: nice, but very far from essential. The authors [of the *Eurobarometer* report] quickly add that 'but this might change in the future if citizens are confronted with serious security problems'. They are right, but the point remains that, at the moment, all the fuss about privacy in the EU is a political rather than a social priority. […] Perhaps we 'do not get it' when we should (a bit like the environmental issues) and need to be better informed. Or perhaps we are informed and still think that other issues are much more pressing.' (500)

This book emphasises the first-mentioned option, i.e. the lack of information and formative discourse. It stresses, moreover, that this notably applies to the disregarded ethical issues and wider societal implications of techno-social big data entanglements. For instance, as long as it remains underemphasised and unclear what ramifications a lack of data protection may have for public health and individual healthcare, important arguments needed for formative discourse are systematically excluded. From a Habermasian perspective, this is less an issue of 'not getting it', but rather a matter of shaping individuals' chances for appreciating an issue and voicing (dis-)approval.

In this context, interdependencies between science, public trust, societal hopes and expectations are of key importance. Van Dijck's work pointedly highlights the relevance of scientists as key pillars of social trust, its formation and mobilisation: 'a paradigm resting on the pillars of academic institutions often forms an arbiter of what counts as fact or opinion, as fact or projection' (2014, 206). In this sense, scientists involved in big data-driven research lend credibility to the assumption that corporate tech data can make a much needed contribution to societal wellbeing, thus potentially justifying compromises regarding individual rights. They give credibility to the (questionable) assumption that corporate data collection approaches are morally indisputable and ethical debates hence unnecessary.

This likewise discourages public negotiations of big data practices, and impedes discursive conditions for which the 'force of the better argument' (Keulartz et al. 2004, 19) is decisive. A major reason for this is that criticism is implicitly framed as unnecessary and futile, as well as selfish and detrimental: unnecessary, since big data's use in public health research asserts the moral appropriateness of corporate data retrieval; futile, since these approaches are authoritatively presented as already established technological and moral 'state of the art'; and selfish and detrimental, considering normative claims for the societal benefits attributed to big data.

Therefore, discursive conditions for big data-driven health research and related norms urgently require amplified, research-driven efforts for facilitating public debate, and the involvement of affected individuals. Yet instead we are witnessing another instance and variation of the pacing problem (Marchant, Allenby and Herkert 2011). While technological innovation has been embraced in big data-driven public health research, scrutinising ethical issues has been largely eschewed, and learning from controversies hindered.

Validity claims

The involvement of data subjects is largely missing in ethical negotiations concerning big data-driven health research. However, normative arguments are brought forward by academics involved in or affected by such research. These discourses illustrate the validity claims through which big data-driven approaches are justified or opposed.

Scholars such as Rothstein and Shoben (2013) as well as Ioannidis (2013) vehemently oppose the argument that informed consent has become irrelevant in big data-driven research. In terms of validity claims, they reject this tendency by raising doubt as to the normative rightness as well as the accuracy of statements made by proponents of big data research. According to the authors, neglecting informed consent neither warrants the alleged methodological advantages, such as the avoidance of (consent) bias nor sufficiently address moral concerns such as the lack of attention to individuals' autonomy and privacy. The latter argument also refers to the conditions of corporate data retrieval. Abandoning informed consent for big data research is seen as potentially hazardous to the reputation of academic research, in particular with regards to public trust. Such arguments brought forward in response to big data-driven research indicate interdependencies between claims presented as part of different discursive domains: 'strictly' moral assumptions and the technological promises of big data can barely be treated separately from each other.

Validity claims to normative rightness (moral justice) as well as validity claims to truth (the factual accuracy of statements) need to be understood as co-constitutive in projects using biomedical big data for public health surveillance. Researchers particularly highlight societal benefits and future possibilities, from *normative* perspectives. They articulate claims to normative rightness, for example in terms of the desirability and expected benefits such as improved public health or cost effectiveness. But these claims to normative rightness are contingent on validity claims to truth, for example with regards to methodological conclusiveness and technological developments.

When considering the use of their data, individuals need to assess whether a certain claim to normative rightness, such as the safeguarding of privacy, may be seen as valid. Likewise, they need insights into the conditions and consequences proposed in related claims to truth: for instance, if the level of privacy proposed as morally reasonable can be indeed safeguarded by certain technologies and methodologies. It is therefore misleading to completely separate statements regarding a technology's functional aspects from normative claims. Along these lines, Swierstra and Rip (2007, 7)[92] even suggest that ultimately, all arguments brought forward in debates on new and emerging technologies are ethical.

In this sense, there is no difference between the ethical, legal, and social aspects (ELSA) in science and technology developments. Instead, '[p]resumably 'non-ethical' arguments in the end refer to stakeholders' interests/rights and/or conceptions of the good life – thus, ethics' (Swierstra and Rip 2007, 7). Swiertsra and Rip stress that this notably applies to discourses on health and environmental risks, which are commonly, yet misleadingly, framed as mainly technological issues. In contrast, the authors emphasise links between technical and ethical matters, reasoning that '[…] the technical discussion can be opened up again to ethical discussion when the assumptions protecting the technical approach are questioned' (ibid.). Bringing this back to Habermas' emphasis on valid social norms as *just* norms, this means that in big data-driven health too, surveillance validity claims to truth and rightness *alike* amount to matters of social justice.

Therefore, to assess the moral reasoning of big data-driven research, we likewise require transparency in terms of methodological and technological conditions. The tech-methodological blackboxing, which is characteristic of big data-driven research, however, obstructs individuals' possibilities to engage with validity claims to truth. The argument above also implies that realistic deliberations regarding big data's contribution to public health are ultimately ethical matters. A main reason for this is that articulated techno-social benefits are commonly mobilised to downplay concerns regarding civic, individual rights. These interdependencies are particularly relevant when considering the institutional conditions of big data-driven health research and its ethics.

Ethical (self-)assessment tends to be constructed as a 'protectionist hurdle': an obstacle to overcome, for example during the grant application process as well as at certain points throughout a study. Once a tech-oriented project has received the approval of the relevant Institutional/Ethics Review Board, or a comparable committee, there are few incentives to engage with ethical issues. For scientists involved in big data-driven research, continuous overtly critical, tech-methodological as well as ethical concerns are unlikely subjects to foreground. They are mostly incentivised to *justify* rather than question their innovation under competitive conditions.

Research projects commonly need to be presented in ways that enable scholars to acquire funding and to publish refereed papers. This leaves little leeway for stressing risks and uncertainties which could undermine a project's feasibility and competitiveness. In the context of big data-driven biomedical research, this has likely facilitated the tendency that contributions to the public good are commonly foregrounded, while ethico-methodological uncertainties are deemphasised. These dynamics also reflect more general insights into novel technosciences, as observed by Rip: 'Newly emerging sciences and technologies live on promises, on the anticipation of a bright future thanks to the emerging technology [...]' (2013, 196). In contrast, foregrounding ethical concerns may challenge the acceptance of innovations and undermine possibilities for funding in tech-centric grant schemes. This also raises the issue that funding programmes need to open up further possibilities for critical engagement with ethical issues.

Facilitated by the abovementioned factors, risks and ethical uncertainties tend to be deemphasised in comparison to benefits for the common good. Issues such as informed consent, privacy, anonymisation, research transparency and methodological sustainability, as well as entanglements between scholarly research and corporate data economies, are at best mentioned, but rarely scrutinised in ethical accounts of big data-driven research. With regards to privacy, scientists indicate that users' current legal rights and laws relevant to corporate data retrieval are decisive for their methodological choices. But critical research indicates that users' privacy *expectations* diverge from current possibilities for privacy management. Moreover, users' current rights and corporate responsibilities remain to be redefined in emerging legal frameworks and data protection policies.

By using, for example, social media data, researchers endorse their collection as morally reasonable. They foster the perception of such data retrieval as the undisputable status quo and the (future) way to go. This is especially problematic when considering the as yet meagre attention paid to potential ethical issues concerning the role of internet and tech corporations. In a call for essays titled 'Fresh Territory for Bioethics: Silicon Valley', Gilbert (on behalf of The Hastings Center) observes that:

> Biomedical researchers are increasingly looking to Silicon Valley for access to human subjects, and Silicon Valley is looking to biomedical researchers for new ventures. These relationships could be a boon to medicine, but they also raise questions about how well-informed the consent process is and how securely the privacy of the subjects' identity and data is kept. Other than a few quotes in the popular press, bioethicists have had little to say on the topic, although those whom I have spoken with agree that more attention is warranted. (2015)[93]

Moral uncertainties and controversial issues, if at all, mainly appear as sidenotes in big data-driven research. Those few researchers investigating ethical issues are often not directly involved in big data-driven research per se. This tendency speaks further to the juxtaposition of, rather than collaboration between, big data scientists and ethicists. Relating this back to the stakeholder constellations, this also means that there is not only little public discursive engagement: in addition, there is a lack of discursive interaction between scholars using big data for health research and those examining such approaches.

From Data-Driven to Data-Discursive Research

Ethical foresight has been emphasised as an indispensable feature of research involving new and emerging technologies (Floridi 2014; Brey, 2012; Einsiedel 2009). Grappling with ethical issues, risks and uncertainties should not be an approach taken in retrospect. Instead ethics should be an integral part of policy-making, regulatory decisions and developments (Floridi 2014, 501). It is characteristic for technological and scientific innovation, however, to move beyond the imaginaries developed in policy-making contexts. Before novel, ethical issues are negotiated in policy-making and governmental regulations, they may have unfolded in research or development phases already, as also implied in the pacing problem. This issue likewise applies to big data and their use in public health surveillance/research.

Therefore, ethical foresight should not be understood merely as a feature of regulatory practices (see also Swierstra and Rip 2007, 17). It is just as relevant to exploratory stages concerning new and emerging technologies, particularly with regards to their role in research. Ethical issues should be foregrounded and debated continuously, but they are often rather reluctantly taken up. Part of the issue is that the work of ethicists is often understood as the opposite of innovation. In contrast, a pragmatist approach to ethics emphasises that moralities are likewise evolving in interaction with technological transformations, among other factors.

Given that valid social norms and ethics require formative discourse, we urgently need a shift from big data-driven to data-discursive approaches in research. What is currently neglected are inclusive, ethical debates on how the morals and norms pertinent to big data practices and particularly research are formed and justified: how are they developing and how should they develop? Whose positions are (not) reflected in these norms? This is also related to the more practical lack of consideration for how big data practices undermine prior modes of discursive involvement: is it ethically reasonable to abandon informed consent in certain studies and, if so, how can these studies provide novel ways to compensate for this?

From a discourse ethics perspective, this also means that research involving big data currently relies on norms whose validity is largely speculative with regards to the (dis-)approval of affected individuals. I therefore argue that researchers need to move away from big data-*driven* approaches, focused merely on techno-methodological innovation, towards data-*discursive* research foregrounding ethical controversies and risks as well as moral change. This discursive development needs to occur in combination with innovative approaches for engaging potentially affected individuals and stakeholders.

Wide, controversial negotiations of ethical decisions and moral principles are crucial for enhancing the validity of social norms. As already indicated above in relation to the conceptualisation of ethics as a field of innovation, such negotiations are considered to be constructive. Or, as Swierstra and Rip (2007) put it in emphasising the relevance of learning and discursive struggle: 'Since Machiavelli, political theorists have pointed out that struggle among an irreducible plurality of perspectives can be productive.' (19) When acknowledging the merit of struggle and controversy, the question arises how to encourage such dynamics and relevant debates.

First, a part of the answer lies in a point stressed above: ethical issues, risks, and contested moralities should not be downplayed, but foregrounded and made accessible to affected individuals in comprehensible ways. This demand of course invites criticism, as being utopian, not least because it conflicts with how academic funding and publication environments commonly function. Such a potential objection, though, highlights the relevance of research funding/grant schemes which do not treat ethical questions as a side-issue of emerging techno-sciences, but as core contributions and the path to innovation.

Second, the abovementioned question indicates the – of course already much debated – relevance of strategies for public engagement and participatory research approaches regarding new techno-scientific developments (see e.g. Pybus, Coté and Blanke 2015, 4; Moser 2014; Rowe and Frewer 2005; Wilsdon and Willis 2004). Within this domain, it also implies that there are certain kinds of debate and involvement which researchers should seek: with regards to health research involving big data, particularly ethical controversies, risks, and changing moralities. The engagement of potentially affected individuals in formative discourses facilitates valid, just norms crucial to emerging forms of public health surveillance using big data.

Stakeholders' involvement and interaction amount to learning processes that have been described as productive struggle (Swierstra and Rip 2007). This emphasis on learning also points to the relevance of notions such as data literacy and (digital) information literacy. Such terms refer partly to the capacity of individuals to contextualise, process, and critically assess data and information which they encounter online (see e.g. Herzog 2015). According to Pybus, Coté and Blanke (2015), '[d]ata literacy can act as an extension and updating of traditional discourses around media literacy by refocusing our attention to the material conditions that surround a user's data within highly proprietary digitised environments' (4). However, they also point to the changing, precarious conditions under which researchers have come to access and handle big data (Haendel, Vasilevsky and Wirz 2012).

Data literacy is just as much a matter of technical expertise as of possibilities for discursive engagement and ethical debate. The importance of involving affected individuals also implies an understanding of data literacy as expertise and engagement which is distributed among multiple stakeholders. The above-mentioned lack of attention for contested moralities and norms in public health research involving big data highlights an urgent need for discussion of the ethical dimensions of data literacy. This applies to the ethical expertise invested in research projects as well as individuals' possibilities for realising, opposing or endorsing the use of their data on moral grounds. In this context, the concept of data literacy is not merely meant to imply users' capacities and responsibility to understand the employment of their data. Instead, it aims at stressing the need for an expertise in and sensibility towards issues beyond practicability and optimisation on the part of data collecting and utilising actors.

Data literacy is not simply a skill which corporations or researchers can demand from the public. Instead, they need to consider, and improve, how they play a part in its formation. Relevant knowledge and skills concerning the implications of new technologies, for example regarding the ramifications for individuals' autonomy, need to be acquired. For this process, public debate, controversy and struggle are crucial. As stakeholders in these debates and dynamics, potentially affected actors should not be simply seen as an obscure public that merely needs to be informed in order to be empowered. Instead, potentially affected individuals need fair chances and opportunities for realising and negotiating research practices which concern rights, risks, uncertainties and moral values. These negotiations may just as much result in approval as in disapproval of norms applicable to current big data. Yet this is a decision which needs to be worked towards by involving relevant stakeholders and creating possibilities for civic debate and engagement.

This demand stands in contrast to current tendencies in big data-driven studies that foster further alienation between researchers and those individuals generating data in the first place. With internet and tech corporations incentivising big data-driven research by offering data or funding, researchers need to account for interdependencies between corporate interests, research developments and ethics. To move towards valid social norms concerning the use of health-indicative big data, scholars need to treat and discuss these data not merely as a technologically enabled opportunity. Instead, they need to be foregrounded as matters of ethics and social justice.

Notes

1 While the term 'data' is treated as a mass noun with a singular verb in this common phrase, it will be treated as a plural in this book.

2 In terms of a definition of 'data', Kitchin points out that data may be understood in various ways, e.g. as commodity, social construct, or public good: 'For example, Floridi (2008) explains that from an epistemological position data are collections of facts, from an informational position data are information, from a computational position data are collections of binary elements that can be processed and transmitted electronically [...].' (Kitchin, 2014a, 4) Similarly, the socio-cultural significance of data may also differ among different actors relating to the same data. Data can 'mean' different things to them: for a user, data collected with a personal wearable device may be a way to realise a healthier lifestyle. For a company such as a supermarket, these data mean a possibility to advertise e.g. dietary products which are more likely to sell to this customer. For a health insurance, the use of a wearable – or the lack thereof – could potentially be an approach to different pricings (of course, this also has then an effect on the meanings of these data for the user).

3 After having to discontinue selling a do-it-yourself health kit for their 'Personal Genome Service' in the US (it can be ordered in the UK though), they are now still offering a general 'Health and Ancestry' service.

4 See e.g. Fuchs, 2014, 167ff. on Facebook's data collection and 131ff. on Google's data-centric capital accumulation.

5   It is unquestionable that we are witnessing fundamental changes in digital
    data collection and the scope of datasets; however, the very quantification
    of such data and their growth turns out to be a challenge. As Kitchin and
    Lauriault (2014) state: 'While there are varying estimates, depending on
    the methodology used, as to the growth of data production caused by big
    data[…], it is clear that there has been a recent step-change in the volume
    of data generated, especially since the start of the new millennium.' (2)

6   At the same time, rhetorics of 'data deluge', 'data tsunami' or 'data explosion'
    also evoke threats and big data's 'potential to create chaos and loss of con-
    trol' (Lupton, 2014b, 107).

7   See also https://wiki.digitalmethods.net/Dmi/DmiAbout.

8   With regards to this tendency, Neff et al. (2017) in turn raise the ques-
    tion 'What would data science look like if its key critics were engaged to
    help improve it, and how might critiques of data science improve with an
    approach that considers the day-to-day practices of data science?' (85).
    Though one can surely ascertain that CDS scholars are already engaged
    in constructive criticism aimed at improvement, this question should be
    understood in the context of calls for an enhanced empirical engagement.
    In light of such contributions, it is also likely that the field of critical data
    studies will develop towards approaches being more actively involved in/
    co-present during data science practices.

9   See also Berry (2011, 11-12).

10 This book pursues a critical data studies perspective focused on research
    on big data. I will not conduct research *with* big data myself. However,
    as explained before, such an exploration is a possible approach to critical
    data studies.

11 'The former treats technology as subservient to values established in other
    social spheres (e.g., politics or culture), while the latter attributes an autono-
    mous cultural force to technology that overrides all traditional or compet-
    ing values.' (Feenberg 2002, 5)

12 This is on the one hand related to concern among CDS scholars that the
    conditions under which such (big) data are generated are problematic. On
    the other hand, also scholars involved in reflective applications of digital
    methods have raised the issue that corporate data sets are increasingly inac-
    cessible (see e.g. Rieder 2016, May 27).

13 This topic was also discussed during two fascinating roundtables on 'Femi-
    nist Big Data' during the Association of Internet Researchers Conference
    2016 in Berlin.

14 See Keulartz et al. (2002), LaFollette (2000), and Joas (1993) for a broader
    contextualisation of pragmatist ethics.

15 Like many other articles on the benefits of activity trackers, also the latter
    webpage 'conveniently' includes a *FitBit* advertisement, referring the user
    to *Amazon*.

16 Rip claims however that his concept addresses a potential shortcoming of pragmatist ethics: While he acknowledges that pragmatist ethics can shed light on normativities which are involved in the development and negotiation of emerging technologies, he criticises the approach's micro-level focus (Rip, 2013, 205ff.).

17 The co-authors are Maartje Schermer, Michiel Korthals and Tsjalling Swierstra.

18 See also Keulartz et al. (2002), LaFollette (2000), Joas (1993) for a wider contextualisation of pragmatist ethics.

19 To further stress the connection with foregoing reflections on *critical* approaches: Hansen suggests that '[…] the critical field, nowadays, is divided between those who hold fast to older paradigms (whether a norm-infused critical theory) or the antifoundationalism of postmodernism and poststructuralism, and those who, believing all foundationalist metanarratives to be false, have opted for a neopragmatic solution, the revamped, ameliorated version of an older pragmatism that proved blind to regimes of power.' (Hansen 2014, 12–13; see also Fraser 1995).

20 In fact, the authors assert that this non-normative insistence has become even more pronounced with the transition from moderate to radical constructivism taken on by some scholars in this field (Keulartz et al. 2004, 13).

21 See Mingers/Walsham (2010) on the relevance of discourse ethics for emerging technologies.

22 Habermas formulates these principles as follows: (U) A valid norm presupposes that '*[a]ll* affected can accept the consequences and the side effects its *general* observance can be anticipated to have for the satisfaction of *everyone's* interests (and these consequences are preferred to those of known alternative possibilities for regulation)' (1990, 65); (D) 'Only those norms can claim to be valid that meet (or could meet) with the approval of all affected in their capacity *as participants in a practical discourse*' (1990, 66). Some authors have criticised that the 'U' principle is redundant (Benhabib 1992, 37ff.) and epistemically problematic (Gottschalk-Mazouz 2000, 186).

23 According to the author, '[j]ust as descriptive statements can be true, and thus express what is the case, so too normative statements can be right and express what has to be done' (Habermas 2001 [1993], 152).

24 Despite this emphasis, when analysing my case studies, I will also refer back to validity claims regarding truth because – as I will show – these are often closely related to arguments concerned with normative rightness.

25 In the translator's introduction to *Justification and Application* (2001), Cronin states that '[y]et, the notion of consensus under ideal conditions of discourse is not an empty ideal without relation to real discursive practices. Habermas maintains that the ideal has concrete practical implications because, insofar as participants in real discourses understand themselves to be engaging in a cooperative search for truth or rightness solely on the basis

of good reasons, they must, as a condition of the intelligibility of the activity they are engaged in, assume that the conditions of the ideal speech situation are satisfied to a sufficient degree' (xv).

26 It has been criticised however, for example with regards to medical, interventional research such as clinical trials in the EU, that there is an urgent need for regulations '[…] improving the widely varying quality of the EU's ethics committees by setting clear quality standards. Leaving these committees just as diverse as before means that European citizens of different member states cannot rely on the same level of protection' (Westra et al. 2014).

27 Recently, some companies – such as the *Google* subsidiary *DeepMind Technologies Ltd.* which is specialised in artificial intelligence research – have launched initiatives aimed at countering this deficit ('DeepMind: Ethics and Society' 2017).

28 When speaking of 'big data', one typically refers to quantified information allowing for insights into individuals' digital as well as physical behaviour, features, or interests. Sensor technologies, for example, likewise produce vast amounts of data concerning the natural world/non-human actors (Hampton et al. 2013). Yet, in this sub-chapter I will focus on data concerning individuals.

29 See also Mittelstadt and Floridi (2016) on the risk of ignoring group-level ethical harms in biomedical research.

30 'The notion of the common good is therefore primarily concerned with the needs and interests of society as a whole, not with individual persons, their interests, and needs.' (Hoedemaekers, Gordijn and Pijnenburg 2006, 417) For a differentiation between public and common good and the implications of these terms see Bialobrzeski, Ried and Dabrock (2012).

31 Likewise, boyd and Crawford suggest that '[t]he current ecosystem around Big Data creates a new kind of digital divide' (2012, 674).

32 In contrast to boyd and Crawford who speak of data rich and data poor groups in order to distinguish between corporate and civic actors/independent scholars, Avital et al. propose in an (earlier) publication to use the term 'data rich' when referring to research teams collecting, analysing, and offering public access to large datasets collaboratively (2007, 4).

33 Even though the image messaging service has been popularised by stressing the (alleged) ephemerality of exchanged content, the mobile app includes advertisement derived from users' communication. These are for example derived from exchanged images and displayed objects that may allow for insights into users' product interests (Vincent 2016).

34 Twitter's data analytics service *Gnip* does not speak of 'big data', but 'social data'.

35 This is occasionally also described as 'data liberation' (Kshetri 2016, 50). This term is likewise used by *Google Inc.*'s so-called 'Data liberation front' (Fitzpatrick 2009) in order to describe their support of users in exporting personal data after opting out of services.

36 Relating this back to the prior section on data asymmetries, it also seems symptomatic that the emotional contagion experiment also triggered some reactions pointing out that overt criticism on this study might have a chilling effect on Facebook's willingness to provide access to data: '[…] haranguing Facebook and other companies like it for publicly disclosing scientifically interesting results of experiments that it is already constantly conducting anyway – and that are directly responsible for many of the *positive* aspects of the user experience – is not likely to accomplish anything useful. If anything, it'll only ensure that, going forward, all of Facebook's societally relevant experimental research is done in the dark, where nobody outside the company can ever find out–or complain–about it.' (Yarkoni 2014; later on, the author commented on his position: see Yarkoni 2014a).

37 'Consent bias' is a type of selection bias which implies that those individuals granting consent differ from those who deny consent, in consequence distorting a study's results.

38 Interestingly, a study on this issue has been published by *Facebook*-affiliated researchers (Bakshy, Messing, and Adamic 2015; see also Pariser 2015 for further reflections on the implications of these insights for the 'filter bubble' theory).

39 In 1999, some of the AOL volunteers filed a class action lawsuit, asking to be compensated for their work based on grounds that it was seen as equivalent to non-paid employment (Hallissey et al. v. America Online). In May 2010, a settlement between AOL and its former volunteers was reached which reportedly resulted in an AOL payment of $15 million (Kirchner 2011).

40 The term was originally coined with regards to pre-digital practices by Toffler (1970).

41 In their article, the authors examine an understanding of biobanks as public or common good and ethical implications for e.g. informed consent; see also Ioannidis 2013: 40ff.

42 In addition, this also seems appropriate given the range of more traditional biomedical big data: 'Biomedical data can take a wide array of forms, ranging from free text (eg, natural language clinical notes) to structured information (eg, such as discharge databases) to high-dimensional information (eg, genome-wide scans of single nucleotide polymorphisms).' (Malin, El Emam and O'Keefe 2013, 4)

43 The directorate itself is of course also a stakeholder.

44 This overview obviously simplifies the stakeholder constellations relevant to big data-driven health research. To briefly address just a few of the stakeholders neglected in this overview: Public health institutions are important entities, because they run public health monitoring programmes and offer related services drawing on more 'traditional' biomedical data (Paul et al. 2016, 470). They are on the one hand significant as existing service providers which may contribute to and may be complemented by big data-driven platforms. On the other hand, public health professionals also act as potential customers of

novel approaches and as important partners for collaboration. For policy-makers, big data-driven emerging approaches open up new, allegedly more cost-effective, efficient ways for public health monitoring and interventions; at the same time, they pose risks which may require regulation. Also those stakeholders that may put the research results of big data-driven studies to practical, potentially economically motivated use have not been covered, for instance, corporate, private/privatised or governmental health insurance providers. In research projects, biomedical big data might be used to gain insights into a population's or person's health condition, self-care or risk behaviour. But the interpretation and use of these data in different context will differ.

[45] On a broader scale, these developments are further substantiated and encouraged by the US. *Big Data Research and Development Initiative* (Executive Office of the President 2016).

[46] According to Weindling (1993): 'Systems of training along American lines were intended to replace the dependence on German and Austrian medical education and public health facilities that had prevailed before 1914. [...] The policy of the Rockefeller Foundation was to transplant American models of public health – in the belief that European institutions had become backward in the course of the war, and that the old guard of professors were militaristic nationalists.' (255)

[47] Together with Google co-founder Sergey Brin, whose wife is one of *23andMe*'s cofounders, GV gave $3.9 in 2007 (series A).

[48] See https://www.google.org/crisisresponse/about.

[49] See http://www.gatesfoundation.org.

[50] See https://givingpledge.org.

[51] I have published parts of this overview in earlier papers (see Richterich 2018; 2016).

[52] See http://websenti.b3e.jussieu.fr/sentiweb/index.php?rub=61.

[53] See http://www.who.int/csr/don/en/index.html.

[54] See http://ecdc.europa.eu/en/press/epidemiological_updates/Pages/epidemiological_updates.aspx.

[55] See http://www.who.int/csr/en/.

[56] The website is not accessible anymore, but has been documented on: http://davidrothman.net/2007/02/22/healthmap-epispider.

[57] A broader overview of platforms and services has been provided by the *SGTF* (see http://endingpandemics.org/community).

[58] The project received initial funding in 2015 and again in 2016, see https://projectreporter.nih.gov/project_info_description.cfm?aid=9146666&icde=31609271.

[59] The project was funded in 2016 with, see https://projectreporter.nih.gov/project_info_description.cfm?aid=9317061&icde=31609271. Young, the PI of this project, is likewise involved in Wang n.d.; both projects are located at the University of California, LA, and closely interrelated.

⁶⁰ See https://www.socialactionlab.org/hsmg; the project was not funded as part of *BD2K*, but by *The National Institute of Allergy and Infectious Diseases* (in total : $450,000).

⁶¹ It received funding for 4 sub-projects from 2013 until 2016 (in total : $750,161); see https://projectreporter.nih.gov/project_info_description.cfm?aid=9127812&icde=31609271.

⁶² While the research by Young, Yu and Wang (2017) particularly investigates posts as units indicating drug and sex-related risk behaviour, other studies have explored the significance of so-called 'action words'. For example, Ireland et al. (2015) argue that verbs, such as go, act, engage etc. '[…] may have reciprocal associations with the level of political activity in a community, and such mobilization has been shown to be associated with reduced HIV rates' (1257). According to the authors, this study was particularly aimed at identifying communities with a heightened HIV risks to support targeted, preventive measures and community support.

⁶³ Trained as a biostatistician, Taha Kass-Hout was the first 'Chief Health Informatics Officer' at the US *Food and Drug Administration*. Among other things, he was involved in *InSTEDD*, founded by *Google, Inc.* in 2006, a social network specialised in early disaster warning and response; moreover, he filed a patent application (together with Massimo Mirabito) for a 'Global disease surveillance platform, and corresponding system and method' (see https://www.google.com/patents/US20090319295).

⁶⁴ Zimmer currently continues his work on big data, social media and internet ethics within the US *National Science Foundation*-funded research team/project *Pervasive Data Ethics for computational research*. While the project pursues broader objectives, it targets many of the issues highlighted with regards to big data-driven health research too.

⁶⁵ Since I have emphasised the role of Internet and tech corporations before, it should be at least pointed out that also in this context such leading companies are not entirely detached: two of the directors, danah boyd and Kate Crawford, are employees of *Microsoft Research*.

⁶⁶ See also Paul et al. (2016).

⁶⁷ Out of the six people who were involved in this paper, five were *Google Inc.* employees.

⁶⁸ Glimpses into *Google*'s big data practices are offered in *Google Trends*, but it is unclear how these vague indicators of search trends have been curated and potentially modified prior to their publication.

⁶⁹ Meanwhile, very practical information supporting related practices have been provided by e.g. marketing agencies (see Points Group n.d. on 'Investing in Google Ads or Facebook Ads for Healthcare').

⁷⁰ This is specified in the 'Competing interests' declaration: 'Dr. Ayers declares that a company he holds equity in has advised the Hopkins Weight Management Center on advertising. However, this does not alter the authors'

adherence to all the PLOS ONE policies on sharing data and materials'. (Chunara et al. 2013, 1)

71 At *The Market Research Event* in Nashville (Tennessee) 2013, their study received the *EXPLOR Award* which 'recognizes breakthrough innovation in technology as applied to Market Research' (see http://www.mktginc.com/thefacts.aspx?service=award).

72 See https://en-gb.facebook.com/business/products/ads/ad-targeting (accessed in February 2017)

73 This category is offered by *Facebook Inc.* as part of their targeted advertising solution called 'US Hispanic Affinity Audience'. As stated on their website: 'The US Hispanic cluster is not designed to identify people who are ethnically Hispanic. It is based on actual users who are interested in or will respond well to Hispanic content, based on how they use Facebook and what they share on Facebook.' (see https://www.facebook.com/business/a/us-hispanic-affinity-audience; accessed in February 2017).

74 See also Giglietto, Rossiand and Bennato (2012) on the '[…] need for collaboration among scientists coming from different backgrounds in order to support studies that combine broad [quantitative] perspectives with in-depth and effective interpretations' (155).

75 'Reactions' was promoted as design choice enabling users to indicate more specific emotions regarding content; shortly after its release though, the function has been described as attempt to create more fine-grained data (Chowdhry 2016).

76 That *Facebook* data are highly sensitive (i.a. health relevant) has also been demonstrated by Kosinskia, Stillwella and Graepel (2013). By implementing an app for involving users in voluntary, consented submissions of personal *Facebook* data, the authors demonstrated that Facebook 'Likes' allow for insights into users' 'sexual orientation, ethnicity, religious and political views, personality traits, intelligence, happiness, use of addictive substances, parental separation, age, and gender' (5802; see also http://mypersonality.org/wiki/doku.php).

77 It has been suggested that some of the European developments can be seen as a 'whiplash effect' responding to prior negligence of ethical issues. Mittelstadt and Floridi state that in such cases it may occur that '[…] overly restrictive measures (especially legislation and policies) are proposed in reaction to perceived harms, which overreact in order to re-establish the primacy of threatened values, such as privacy. Such a situation may be occurring at present as reflected in the debate on the proposed European Data Protection Regulation currently under consideration by the European Parliament (Wellcome Trust 2013), which may drastically restrict information-based medical research utilising aggregated datasets to uphold to uphold ethical ideals of data protection and informed consent' (2015, 305; see also chapter 3 on informed consent).

[78] The authors likewise point out that '[…] from the Facebook Advertisement platform, there are no metrics provided about precision of estimates, however the information is freely available, and resolution was accurate enough for the purposes of this study.' (Chunara et al. 2012, 6)

[79] Van Dijck furthermore addresses the normative implications, context dependences, and risks of future data utilisation: 'Messages from millions of female Facebook users between 25 and 35 years of age posting baby pictures in their Timeline may be endlessly scrutinised for behavioural, medical, or consumerist patterns. Do researchers want to learn about young mothers' dieting habits with the intention of inserting propositions to change lifestyles? Or do they want to discover patterns of consumptive needs in order for companies to sell baby products at exactly the right moment? Or, perhaps far-fetched, are government agencies interested in interpreting these data for signs of postnatal depression or potential future child abuse?' (2014, 202).

[80] See http://www.unglobalpulse.org/projects/haze-gazer-a-crisis-analysis-tool (accessed February 2017).

[81] See http://www.unglobalpulse.org/projects/monitoring-hiv-mother-child-prevention-programme (accessed February 2017).

[82] GFT has been discontinued as public service in 2015 (Lazer and Kennedy 2015).

[83] Parts of the analyses were already published in Richterich (2018). I discuss this particular example in this book too, since it is a materialisation of research from a team which has been extensively engaged in exploring the possibilities for big data-driven public health surveillance. Moreover, the application has received some (mainly positive) media attention: On March 14, the site selected and mapped a report on a 'mysterious hemorrhagic fever' killing 8 people in Guinea (Asokan and Asokan, 2015; Kotz 2015). Only several days later, on March 23, the *WHO* published a first official report on the Ebola outbreak.

[84] These twelve categories are: Animal, Environmental, Fever/Febrile, Gastro-intestinal, Hemorrhagic, Hospital Acquired Infection, Neurologic, Other, Respiratory, Skin/Rash, STD, and Vectorborne. These are represented to a different extent in the current overview. For example, on 20th July, 2016, out of 889 alerts from the past week, the main part of reports were related to 'Vectorborne Alerts' (742 in total; e.g. Dengue and Zika virus), and 'Respiratory Alerts' (202 in total; e.g. Influenza H1N1 and Tuberculosis). In July 2016, the website put particular emphasis on the 2016 Zika virus pandemic. The virus is dominantly spread by Aedes mosquitos and causes symptoms such as (mild) fever, muscle pain, or headache. Infections are particularly harmful in case of pregnant women, since the virus can cause birth defects.

[85] An overview of sources is also provided on the website's 'About' section: http://www.healthmap.org/site/about. While *Twitter* references can be

found on the map, the microblogging platform is currently (July 2016) not mentioned as source.

[86] One should consider that this post has been published in 2008, just after *HealthMap* received Google.org funding.

[87] The critical debate concerning neogeography is also reflected in the broader wariness of allegedly participatory digital cultures (see e.g. Fuchs 2014: 52ff; Van Dijck 2009).

[88] The author refers back to the work of Vaidhyanathan on *The Googlization of Everything* (2012).

[89] Despite public trust in science having been described as eroding (Wynne 2006, 212ff.), in most 'Western' countries, scientists are ranked among the most trustworthy societal actors (Rathenau Institute 2016; Eurobarometer–419 2014; Castell et al. 2014; Van Dijck 2014; Gauchat 2012).

[90] A strong, causal link between scandals and regulatory developments is however contested. Hedgecoe advises caution in attributing regulatory changes simply to research scandals. He suggests '[...] that changes may be misattributed to external events (shocks, punctuation marks) but that a finer-grained study of organizational history and process reveals an underlying current of change that is driven by quite different forces.' (2016, 591)

[91] Apart from the modes of funding and domains I mentioned especially in chapter 4, others are likewise remarkable: for example, *Google Inc.'s* ad grants programmes offer non-profit institutions up to '$10,000 USD of in-kind advertising every month from *AdWords*' (https://www.google.com/grants). These programmes received attention in the context of recent debates on Google search results with high-ranking Holocaust denial pages: when it became known that the Breman Museum, a Jewish heritage museum in Atlanta, had to rely on such a grant in order to '[…] pay for adverts that counter search results that appear to deny that the Holocaust happened' (Cadwalladr 2016).

[92] The authors likewise draw on a pragmatist approach to ethics, arguing that while general consensus may be impossible to reach, struggle, reflexivity and learning are key benefits of techno-moral negotiations. Similar to Habermas' theory being described as counterfactual idealization, also the authors state that '[…] even if the agora is an illusion, it is a necessary one, and it is productive' (2007, 19).

[93] The call is still ongoing and some of the reactions have been cited in this book (Sharon 2016a; Drabiak 2016).

References

Alba, Davey. 2016. 'Google.org Thinks it can Engineer a Solution to the World's Woes.' *Wired*, August 3. Retrieved from: https://www.wired.com/2016/03/giving-google-way.

Albarracin, Dolores et al. n.d. 'Online Media and Structural Influences on New HIV/STI Cases in the US.' Retrieved from: https://www.socialactionlab.org/hsmg.

Andrejevic, Mark, and Kelly Gates. 2014. 'Editorial. Big Data Surveillance: Introduction.' *Surveillance & Society*, 12(2): 185–196.

Andrejevic, Mark. 2013. 'Estranged Free Labor.' In Trebor Scholz (ed.), *Digital Labor: The Internet as Playground and Factory*, pp. 149–164. Routledge: New York

Andrejevic, Mark. 2014. 'Big Data, Big Questions: The Big Data Divide.' *International Journal of Communication*, 8: 1673–1689.

Annas, George J., and Sherman Elias. 2014. '23andMe and the FDA.' *New England Journal of Medicine*, 370(11): 985–988.

Antheunis, Marjolijn L., Kiek Tates, and Theodoor E. Nieboer. 2013. 'Patients' and Health Professionals' Use of Social Media in Health Care: Motives, Barriers and Expectations.' *Patient Education and Counseling*, 92(3): 426–431.

Apel, Karl-Otto. 1984. *Understanding and Explanation: A Transcendental-Pragmatic Perspective*. Cambridge, Mass: MIT Press.

Apel, Karl-Otto. 1988. *Diskurs und Verantwortung: Das Problem des Übergangs zur postkonventionellen Moral*. Berlin: Suhrkamp.

Arrington, Michael. 2006. 'AOL Proudly Releases Massive Amounts of Private Data.' *TechCrunch,* August 6. Retrieved from: http://techcrunch.com/2006/08/06/aol-proudly-releases-massive-amounts-of-user-search-data.

Ashmore, Malcolm. 1996. 'Ending Up on the Wrong Side: Must the Two Forms of Radicalism Always be at War?' *Social Studies of Science,* 26(2): 305–322.

Assunção, Marcos et al. 2015. Big Data Computing and Clouds: Trends and Future Directions. *Journal of Parallel and Distributed Computing,* 79, 3–15.

Auffray, Charles et al. 2016. 'Making Sense of Big Data in Health Research: Towards an EU Action Plan.' *Genome Medicine,* 8(1). Retrieved from: https://doi.org/10.1186/s13073–016–0323-y.

Avital, Michel et al. 2007. 'Data Rich and Data Poor Scholarship: Where Does IS Research Stand?' *ICIS 2007 Proceedings.* Retrieved from: http://aisel.aisnet.org/icis2007/101.

Baert, Patrick, and Bryan Turner. 2004. 'New Pragmatism and Old Europe: Introduction to the Debate between Pragmatist Philosophy and European Social and Political Theory.' *European Journal of Social Theory,* 7(3): 267–274.

Baeza-Yates, Ricardo. 2016. 'Data and Algorithmic Bias in the Web.' In *Proceedings of the 8th ACM Conference on Web Science.* DOI: https://doi.org/10.1145/2908131.2908135.

Bakshy, Eytan, Solomon Messing, and Lada A. Adamic. 2015. 'Exposure to Ideologically Diverse News and Opinion on Facebook.' *Science,* 348(6239): 1130–1132.

Baruh, Lemi and Mihaela Popescu 2015. 'Big Data Analytics and the Limits of Privacy Self-Management.' *New Media & Society.* Advance online publication. DOI: https://doi.org/10.1177/1461444815614001. 1–18.

Benhabib, Seyla. 1992. *Situating the Self: Gender, Community, and Postmodernism in Contemporary Ethics.* New York: Routledge.

Benner, Mats and Ulf Sandström. 2000. 'Institutionalizing the Triple Helix: Research Funding and Norms in the Academic System.' *Research Policy,* 29(2): 291–301.

Beresford, Alastair R. and Frank Stajano. 2003. 'Location Privacy in Pervasive Computing.' *IEEE Pervasive Computing,* 2(1), 46–55.

Berezin, Alexander. 1998. 'The Perils of Centralized Research Funding Systems.' *Knowledge, Technology & Policy,* 11(3): 5–26.

Bekelman, Justin E., Yan Li and Cary P. Gross. 2003. 'Scope and Impact of Financial Conflicts of Interest in Biomedical Research: A Systematic Review.' *JAMA,* 289(4): 454–465.

Bernstein, Jay M. 2014. *Recovering Ethical Life: Jurgen Habermas and the Future of Critical Theory.* Abingdon: Routledge.

Berry, David 2011. The Computational Turn: Thinking about the Digital Humanities. *Culture Machine.* Retrieved from: http://www.culturemachine.net/index.php/cm/article/view/440/470.

Berry, David. 2011a. *The Philosophy of Software: Code and Mediation in the Digital Age*. Basingstoke: Palgrave.

Bertot, John Carlo et al. 2014. Big Data, Open Government and e-Government: Issues, Policies and Recommendations. *Information Polity*, 19(1, 2), 5–16.

Bialobrzeski, Arndt, Jens Ried, and Peter Dabrock. 2012. 'Differentiating and Evaluating Common Good and Public Good: Making Implicit Assumptions Explicit in the Contexts of Consent and Duty to Participate.' *Public Health Genomics*, 15(5): 285–292.

BigDataEurope. 2016. 'Integrating Big Data, Software and Communities for Addressing Europe's Societal Challenges.' Retrieved from: http://cordis.europa.eu/project/rcn/194216_en.html.

Bijker, Wiebe. 1995. *Bicycles, Bakelites and Bulbs: Toward a Theory of Sociotechnical Change*. Cambridge, MA: MIT Press.

Bloustein, Edward J. 1976. 'Group Privacy: The Right to Huddle.' *Rutgers Law Journal*, 8: 219–283.

Boellstorff, Tom and Maurer, Bill. 2015. 'Introduction.' In Tom Boellstorff and Bill Maurer, eds. *Data, Now Bigger and Better!*, pp. 1-6. Chicago, IL: Prickly Paradigm Press.

Boellstorff, Tom. 2013. 'Making Big Data, in Theory.' *First Monday*, 18(10). Retrieved from: http://firstmonday.org/ojs/index.php/fm/article/view/4869/3750.

Bogost, Ian, and Nick Montfort. 2009. *Racing the Beam: The Atari Video Computer System*. Cambridge, MA: MIT Press.

Booth, Robert. 2014. 'Facebook Reveals News Feed Experiment to Control Emotions.' *The Guardian*, June 30. Retrieved from: https://www.theguardian.com/technology/2014/jun/29/facebook-users-emotions-news-feeds.

Borgman, Christine L. 2015. *Big Data, Little Data, No Data: Scholarship in the Networked World*. Cambridge, Mass: MIT.

Bowker, Geoffrey C. 2013. 'Data Flakes: An Afterword to 'Raw Data' Is An Oxymoron' In Lisa Gitelman, ed, pp. 167–171. *'Raw Data' Is An Oxymoron*. Cambridge, Mass: MIT Press .

Boyd, Andrew. 2017. 'Could Fitbit Data Be Used to Deny Health Coverage?' *US News*, February 17. Retrieved from: https://www.usnews.com/news/national-news/articles/2017–02-17/could-fitbit-data-be-used-to-deny-health-insurance-coverage.

boyd, danah and Kate Crawford. 2012. 'Critical Questions for Big Data: Provocations for a Cultural, Technological, and Scholarly Phenomenon.' *Information, Communication & Society*, 15(5), 662–679.

boyd, danah and Nicole B. Ellison. 2007. 'Social Network Sites: Definition, History, and Scholarship.' *Journal of Computer-Mediated Communication*, 13(1): 210–230.

boyd, danah. 2010. 'Privacy and Publicity in the Context of Big Data.' Talk at WWW. Raleigh, North Carolina, April 29. Retrieved from: www.danah.org/papers/talks/2010/WWW2010.html.

Bozdag, Engin. 2013. 'Bias in Algorithmic Filtering and Personalization.' *Ethics and Information Technology*, 15(3): 209–227.

Brandt, Allan M. and Martha Gardner. 2013. 'The Golden Age of Medicine?' In Roger Cooter and John Pickstone, eds., *Companion Encyclopedia of Medicine in the Twentieth Century*, pp. 21–38. Abingdon and New York: Routledge.

Brey, Philip A. 2012. 'Anticipatory Ethics for Emerging Technologies.' *NanoEthics*, 6(1), 1–13.

Bronson, Kelly, and Irena Knezevic. 2016. 'Big Data in Food and Agriculture.' *Big Data & Society*, 3(1). DOI: https://doi.org/2053951716648174.

Brownstein, John S. et al. 2008. 'Surveillance Sans Frontieres: Internet-Based Emerging Infectious Disease Intelligence and the HealthMap Project.' *PLoS Med*, 5(7). Retrieved from: http://journals.plos.org/plosmedicine/article?id=10.1371/journal.pmed.0050151.

Brownstein, John S., and Clark Freifeld 2007. 'HealthMap: The Development of Automated Real-Time Internet Surveillance for Epidemic Intelligence.' *Euro Surveillance*, 12(11). Retrieved from http://www.eurosurveillance.org/viewarticle.aspx?articleid=3322.

Brownstein, John S., Clark Freifeld, and Lawrence C. Madoff. 2009. 'Digital Disease Detection – Harnessing the Web for Public Health Surveillance.' *New England Journal of Medicine*, 360(21): 2153–2157.

Bundesministerium für Bildung und Forschung. n.d. 'Big Data – Management und Analyse großer Datenmengen.' Retrieved from: https://www.bmbf.de/de/big-data-management-und-analyse-grosser-datenmengen-851.html.

Burgess, Jean and Axel Bruns. 2012. 'Twitter Archives and the Challenges of 'Big Social Data' for Media and Communication Research.' *M/C Journal*, 15(5). Retrieved from: http://journal.media culture.org.au/index.php/mcjournal/article/viewArticle/561Driscoll.

Burns, Ryan. 2015. 'Rethinking Big Data in Digital Humanitarianism: Practices, Epistemologies, and Social Relations.' *GeoJournal*, 80(4), 477–490.

Burton, Betsy. 2015. 'VP Distinguished Analyst Betsy Burton Discusses the Trends in This Year's Hype Cycles.' Retrieved from: http://www.gartner.com/technology/research/hype-cycles.

Butler, Declan. 2013. 'When Google Got Flu Wrong: US Outbreak Foxes a Leading Web-Based Method for Tracking Seasonal Flu.' *Nature*, 494: 155–156.

Cadwalladr, Carole. 2016. 'Jewish Museum Relies on Google Grant to Counter Holocaust Denial Search Results.' *The Guardian*, December 22. Retrieved from: https://www.theguardian.com/technology/2016/dec/22/google-profiting-holocaust-denial-jewish-breman-museum.

Caspary, William R. 2000. *Dewey on Democracy*. Ithaca: Cornell University Press.

Cassidy, John. 2015. 'Mark Zuckerberg and the Rise of Philanthrocapitalism.' *The New Yorker*, December 2. Retrieved from: http://www.newyorker.com/news/john-cassidy/mark-zuckerberg-and-the-rise-of-philanthrocapitalism.

Castell, Sarah et al. 2014. 'Public Attitudes to Science 2014'. Retrieved from: https://www.britishscienceassociation.org/Handlers/Download.ashx?IDMF=276d302a-5fe8-4fc9-a9a3-26abfab35222.

ChaChaEnterprises, LLC. 2013. 'Case Number: 5:13-bk-53894'. Retrieved from: https://businessbankruptcies.com/cases/cha-cha-enterprises-llc.

Chambers, Chris. 2014. Tweet, June 24. Retrieved from: https://twitter.com/chrisdc77/status/483256460760346624.

Chan, Yu-Feng Yvonne et al. 2017. 'The Asthma Mobile Health Study, a large-scale clinical observational study using ResearchKit'. Nature biotechnology, 35(4): 354-362.

Chang, Lulu. 2016. Wearables are Already Impacting the Healthcare Industry, and Here's How. *Digital Trends*, April 10. Retrieved from: https://www.digitaltrends.com/wearables/wearables-healthcare.

Chen, Sophia. 2017. 'AI Research Is in Desperate Need of an Ethical Watchdog.' *Wired*, September 18. Retrieved from: https://www.wired.com/story/ai-research-is-in-desperate-need-of-an-ethical-watchdog.

Chen, Philip, and Chun-Yang Zhang. 2014. 'Data-Intensive Applications, Challenges, Techniques and Technologies: A Survey on Big Data.' *Information Sciences*, 275: 314–347.

Chiolero, Arnaud. 2013. 'Big Data in Epidemiology: Too Big to Fail?' *Epidemiology*, 24(6): 938–939.

Chowdhry, Amit. 2016. 'Facebook Emoji 'Reactions': Are There Ulterior Motives?' *Forbes*, February 29. Retrieved from: http://www.forbes.com/sites/amitchowdhry/2016/02/29/facebook-reactions/#5dc6c29a7896.

Chunara, Rumi et al. 2012. 'Social and News Media Enable Estimation of Epidemiological Patterns Early in the 2010 Haitian Cholera Outbreak.' *The American Journal of Tropical Medicine and Hygiene*, 86(1): 39–45.

Clarke, Keith et al. 1996. 'On Epidemiology and Geographic Information Systems: A Review and Discussion of Future Directions.' *Emerging Infectious Diseases*, 2(2): 85–92.

Clausen, René N. 2016. 'Celebrating a Decade of Twitter – One of the Largest Platforms for Big Data'. *Pulse Lab Diaries*, March 25. Retrieved from: http://www.unglobalpulse.org/celebrating-decade-twitter-one-largest-platforms-big-data.

Cleveland, William S. 2001. 'Data Science: An Action Plan for Expanding the Technical Areas of the Field of Statistics.' *International Statistical Review*, 69(1); 21–26.

Cohen, Julie E. 2013. 'What Privacy is For.' *Harvard Law Review*, 126(7): 1904–1933.

Collier, Nigel et al. 2008. 'BioCaster: Detecting Public Health Rumors with a Web-based Text Mining System.' *Bioinformatics*, 24: 2940–2941.

Collins, Tom. 2016. 'How Safe is Your Fitness Tracker? Hackers Could Steal Your Data and Sell the Information to Health Companies.' *Daily Mail*, December 19. Retrieved from: http://www.dailymail.co.uk/sciencetech/article-4049154/

How-safe-fitness-tracker-Hackers-steal-data-sell-information-health-companies.html.

COM 442 final. 2014. 'Towards a Thriving Data-Driven Economy.' Retrieved from: http://eur-lex.europa.eu/legal-content/EN/TXT/?qid=1404888011738&uri=CELEX:52014DC0442.

COM 9 final. 2017. 'Building a European Data Economy.' Retrieved from: https://ec.europa.eu/digital-single-market/en/news/communication-building-european-data-economy.

Conway, Michael A. n.d. 'Utilizing Social Media as a Resource for Mental Health Surveillance.' Retrieved from: https://projectreporter.nih.gov/project_info_description.cfm?aid=9127812&icde=31609271.

Cooter, Roger and Pickstone, John, eds., 2013. *Companion Encyclopedia of Medicine in the Twentieth Century*. Abingdon and New York: Routledge.

Costa, Fabricio F. 2014. 'Big Data in Biomedicine.' *Drug Discovery Today*, 19(4): 433–440.

Council, Jared. 2016. 'ChaCha, Unable to Find Financial Answers, Shuts Down Operations Indianapolis.' *Business Journal*, December 12. Retrieved from: http://www.ibj.com/articles/61651-chacha-unable-to-find-financial-answers-shuts-down-operations.

Craig, Terence and Mary E. Ludloff. 2011. *Privacy and Big Data: The Players, Regulators, and Stakeholders*. Beijing: O'Reilly.

Crampton, Jeremy W. 2010. *Mapping: A Critical Introduction to Cartography and GIS*. Chichester: John Wiley & Sons.

Crawford, Kate, Mary L. Gray, and Kate Miltner. 2014. 'Critiquing Big Data: Politics, Ethics, Epistemology: Introduction.' *International Journal of Communication*, 8: 1663–1672

Crawford, Kate, and Jason Schultz. 2014. 'Big Data and Due Process: Toward a Framework to Redress Predictive Privacy Harms.' *Boston College Law Review*, 55(1): 93–128.

Crawford, Kate, and Megan Finn. 2015. 'The Limits of Crisis Data: Analytical and Ethical Challenges of Using Social and Mobile Data to Understand Disasters.' *GeoJournal*, 80(4): 491–502.

Crawford, Kate. 2014. 'Big Data Stalking. Data Brokers Cannot be Trusted to Regulate Themselves.' *Scientific American*, 310(4), 14–14.

Crisp, Arthur H. et al. 2000. 'Stigmatisation of People with Mental Illnesses.' *The British Journal of Psychiatry*, 177(1): 4–7.

Cronin, Ciaran. 2001. 'Translator's Introduction.' In *Justification and Application Remarks on Discourse Ethics* by Jürgen Habermas, (xi–xxix). Cambridge, Mass: MIT Press.

Cuttler, Sasha. 2015. 'San Francisco's General Hospital Belongs to the Public — It's Not Zuckerberg's Plaything.' *Alternet*, December 20. Retrieved from: http://www.alternet.org/personal-health/san-franciscos-general-hospital-belongs-public-its-not-zuckerbergs-plaything.

Dalton, Craig, and Jim Thatcher. 2014. What Does a Critical Data Studies Look Like and Why Do We Care? *Blog post: Society and Space*. Retrieved from: http://societyandspace.com/2014/05/19/dalton-and-thatcher-commentary-what-does-a-critical-data-studies-look-like-and-why-do-we-care.

Dalton, Craig, and Jim Thatcher. 2014. 'Inflated Granularity: The Promise of Big Data and the Need for a Critical Data Studies.' Presentation at the Annual Meeting of the Association of American Geographers, Tampa.

Dalton, Craig M., Linnett Taylor and Jim Thatcher. 2016. 'Critical Data Studies: A Dialog on Data and Space.' *Big Data & Society*, 3(1): 1–9.

Data Science at NIH. 2016. 'Big Data to Knowledge.' Retrieved from: https://datascience.nih.gov/bd2k.

Davenport, Thomas H. and Dhanurjay Patil. 2012. 'Data Scientist: The Sexiest Job of the 21st Century.' *Harvard Business Review*. Retrieved from: https://hbr.org/2012/10/data-scientist-the-sexiest-job-of-the-21st-century .

Deacon, Harriet. 2006. 'Towards a Sustainable Theory of Health-Related Stigma: Lessons from the HIV/AIDS Literature.' *Journal of Community & Applied Social Psychology*, 16(6): 418–425.

Dean, Andrew G. et. al. 1994. 'Computerizing Public Health Surveillance Systems.' In Steven Teutsch and Elliott Churchill, eds., *Principles and Practice of Public Health Surveillance*, pp. 245–270. New York: Oxford University Press.

Dechlich, S. and Anne O. Carter. 1994. 'Public Health Surveillance: Historical Origins, Methods and Evaluation.' *Bulletin of the World Health Organization*, 72: 285–304.

'DeepMind: Ethics and Society.' 2017. Deep Mind, October. Retrieved from: https://deepmind.com/applied/deepmind-ethics-society.

Dencik, Lina, Arne Hintz and Jonathan Cable. 2016. 'Towards Data Justice? The Ambiguity of Anti-Surveillance Resistance in Political Activism.' *Big Data & Society*, 3(2): DOI: https://doi.org/2053951716679678.

Dickstein, Morris, ed. 1998. *The Revival of Pragmatism: New Essays on Social Thought, Law, and Culture*. Durham, NC: Duke University Press.

Douglas, Tyler. 2016. 'The Rise and Fall of Big Data Hype—and What It Means for Customer Intelligence.' Retrieved from: https://www.visioncritical.com/fall-of-big-data-hype.

Drabiak, Katherine. 2016. 'Read the Fine Print Before Sending Your Spit to 23andMe.' *The Hastings Center: Bioethics Forum*. Retrieved from: http://www.thehastingscenter.org/response-to-call-for-essays-read-the-fine-print-before-sending-your-spit-to-23andme-r.

Driscoll, Kevin, and Shawn Walker. 2014. 'Big Data, Big Questions| Working Within a Black Box: Transparency in the Collection and Production of Big Twitter Data.' *International Journal of Communication*, 8: 1745–1764.

Einsiedel, Edna F. ed. 2009. *Emerging Technologies: From Hindsight to Foresight*. Vancouver: UBC.

Eke, Paul I. 2011. 'Using Social Media for Research and Public Health Surveillance.' *Journal of Dental Research*, 90(9): 1045–1046.

Electronic Frontier Foundation. n.d. 'Medical Privacy.' Retrieved from: https://www.eff.org/issues/medical-privacy.

EMC. 2014. Executive Summary. Data Growth, Business Opportunities, and the IT Imperatives. Retrieved from: http://www.emc.com/leadership/digital-universe/2014iview/executive-summary.htm.

'EU and US Strengthen Their Collaboration on eHealth IT'. 2016. *European Commission*, July 7. Retrieved from: https://ec.europa.eu/digital-single-market/en/news/eu-and-us-strengthen-their-collaboration-ehealth-it.

Eurobarometer–419. 2014. 'Public Perception of Science, Research and Innovation'. Retrieved from: http://ec.europa.eu/commfrontoffice/publicopinion/archives/ebs/ebs_419_en.pdf.

European Commission. 2014. 'Factsheet on the 'Right to be Forgotten Ruling''. Available at: http://ec.europa.eu/justice/data-protection/files/factsheets/factsheet_data_protection_en.pdf.

Executive Office of the President: National Science and Technology Council. 2016. 'The Federal Big Data Research and Development Strategic Plan.' Retrieved from: https://www.nitrd.gov/PUBS/bigdatardstrategicplan.pdf.

European Union Agency for Fundamental Right. n.d. 'Information Society, Privacy and Data Protection.' Retrieved from: http://fra.europa.eu/en/theme/information-society-privacy-and-data-protection.

Eysenbach, Gunther. 2002: 'Infodemiology: The Epidemiology of (Mis)information.' *American Journal of Medicine*, 113: 763–765.

Eysenbach, Gunther. 2003. 'SARS and Population Health Technology.' *Journal of Medical Internet Research*, 5(2): https://www.ncbi.nlm.nih.gov/pmc/articles/PMC1550560.

Eysenbach, Gunther. 2006. 'Infodemiology: Tracking Flu-Related Searches on the Web for Syndromic Surveillance.' *AMI1A Annual Symposium*, Proceedings: 244–248.

Eysenbach, Gunther. 2009. 'Infodemiology and Infoveillance: Framework for an Emerging Set of Public Health Informatics Methods to Analyze Search, Communication and Publication Behavior on the Internet.' *Journal of Medical Internet Research*, 11(1): http://www.jmir.org/2009/1/e11.

Faden, Ruth R. and Tom L. Beauchamp. 1986. *A History and Theory of Informed Consent*. Oxford: Oxford University Press.

Farr, Christina and Alexei Oreskovic. 2013. 'Facebook Plots First Steps Into Healthcare.' Reuters, December 21. Retrieved from: https://www.reuters.com/article/us-facebook-health-idUSKCN0HS09720141003?feedType=RSS&feedName=technologyNews.

Finlay, Steven. 2014. *Predictive Analytics, Data Mining and Big Data: Myths, Misconceptions and Methods*. Basingstoke, Hampshire: Palgrave Macmillan.

Fitzpatrick, Brian. 2009. 'Introducing DataLiberation.org: Liberate Your Data!' *Google Public Policy Blog*. Retrieved from: https://publicpolicy.googleblog. com/2009/09/introducing-dataliberationorg-liberate.html.

Floridi, Luciano. 2008. 'Data.' In William Darity, ed. *International Encyclopedia of the Social Sciences*. Detroit: Macmillan.

Floridi, Luciano. 2014. 'Open Data, Data Protection, and Group Privacy.' *Philosophy & Technology*, 27(1): 1–3.

Floridi, Luciano. 2014. 'Technoscience and Ethics Foresight.' *Philosophy & Technology*, 27: 499–501.

Foucault, Michel. 1975. *Discipline and Punish: The Birth of the Prison*. New York: Random House.

Fox, Stephen and Tuan Do. 2013. 'Getting Real About Big Data: Applying Critical Realism to Analyse Big Data Hype.' *International Journal of Managing Projects in Business*, 6(4), 739–760.

Franks, Bill. 2012. *Taming the Big Data Tidal Wave: Finding Opportunities in Huge Data Streams with Advanced Analytics*. Hoboken, NJ.: Wiley.

Fraser, Nancy. 1995. 'Pragmatism, Feminism, and the Linguistic Turn.' In Seyla Benhabib et al., eds., *Feminist Contentions: A Philosophical Exchange* New York: Routledge, pp. 157–171.

Freifeld, Clark C. et al. 2008. 'HealthMap: Global Infectious Disease Monitoring Through Automated Classification and Visualization of Internet Media Reports.' *Journal of the American Medical Informatics Association*, 15(2): 150–157.

Freifeld, Clark C. et al. 2010. 'Participatory Epidemiology: Use of Mobile Phones for Community-Based Health Reporting.' *PLoS Med*, 7(12). Available at: http://journals.plos.org/plosmedicine/article?id=10.1371/journal. pmed.1000376.

Freifeld, Clark C. et al. 2014. 'Digital Drug Safety Surveillance: Monitoring Pharmaceutical Products in Twitter.' *Drug Safety*, 37(5): 343–350.

Friedman, Batya, and Helen Nissenbaum. 1996. 'Bias in Computer Systems.' *ACM Transactions on Information Systems*, 14(3), 330–347.

Fuchs, Christian. 2011. 'Web 2.0, Prosumption, and Surveillance.' *Surveillance & Society*, 8(3): 288–309.

Fuchs, Christian. 2014. *Social Media: A Critical Introduction*. London: Sage.

Fuller, Matthew. 2003. *Behind the Blip: Essays on the Culture of Software*. New York: Autonomedia.

Gandomi, Amir, and Murtaza Haider. 2015. 'Beyond the Hype: Big Data Concepts, Methods, and Analytics.' *International Journal of Information Management*, 35(2), 137–144.

Gauchat, Gordon. 2012. 'Politicization of Science in the Public Sphere: A Study of Public Trust in the United States, 1974 to 2010.' *American Sociological Review*, 77(2): 167–187.

Gerlitz, Carolin and Anne Helmond. 2013. 'The Like Economy: Social Buttons and the Data-Intensive Web.' *New Media & Society*, Advance online publication. DOI: https://doi.org/10.1177/1461444812472322.

Geuna, Aldo. 2001. 'The Changing Rationale for European University Research Funding: Are There Negative Unintended Consequences?' *Journal of Economic Issues*, 35(3): 607–632.

Giannetto, David. 2015. 'The Future of Twitter? Statistics Show 2016 Death'. *Huffington Post*, February 15. Retrieved from: http://www.huffingtonpost. com/david-giannetto/the-future-of-twitter-sta_b_9232280.html.

Gibbs, Samuel. 2014. 'Gmail Does Scan All Emails, New Google Terms Clarify.' *The Guardian*. Retrieved from: http://www.theguardian.com/technology/2014/apr/15/gmail-scans-all-emails-new-google-terms-clarify.

Giglietto, Fabio, Luca Rossi, and Davide Bennato. 2012. 'The Open Laboratory: Limits and Possibilities of Using Facebook, Twitter, and YouTube as a Research Data Source.' *Journal of Technology in Human Services*, 30(3/4): 145–159.

Gilbert, Susan. 2015. 'Fresh Territory for Bioethics: Silicon Valley'. *The Hastings Center: Bioethics Forum*. Retrieved from: http://www.thehastingscenter. org/fresh-territory-for-bioethics-silicon-valley.

Gillespie, Tarleton and Nick Seaver. 2015. 'Critical Algorithm Studies: A Reading List.' Retrieved from: https://socialmediacollective.org/reading-lists/critical-algorithm-studies.

Ginsberg, Jeremy et al. 2009. 'Detecting Influenza Epidemics Using Search Engine Query Data.' *Nature*, 457(7232): 1012–1014.

Gitelman, Lisa. 2013. *'Raw Data' Is an Oxymoron*. Cambridge, Mass: MIT Press.

Glenn, Richard (2003). *The Right to Privacy: Rights and Liberties Under the Law*. Santa Barbara, CA: Abc-Clio.

Godin, Dan. 2013. 'Think your Skype messages get end-to-end encryption? Think again.' *ArsTechnica*. Retrieved from: http://arstechnica.com/security/2013/05/think-your-skype-messages-get-end-to-end-encryption-think-again.

Goffey, Andrew. 2008. 'Algorithm.' In Matthew Fuller, eds., *Software Studies: A Lexicon*. Cambridge, Mass: MIT Press, pp. 15–20.

González Fuster, Gloria (2014). *The Emergence of Personal Data Protection as a Fundamental Right of the EU*. Heidelberg: Springer.

Google Transparency Project. 2017. 'Google Academics Inc.' *Campaign for Accountability*. Retrieved from: http://googletransparencyproject.org/articles/google-academics-inc.

Gottschalk-Mazouz, Niels. 2000. *Diskursethik: Theorien, Entwicklungen, Perspektiven*. Berlin: Akademie.

Gurstein, Michael B. 2011. 'Open Data: Empowering the Empowered or effective Data Use for Everyone?' *First Monday*, 16(2). Retrieved from: http://firstmonday.org/ojs/index.php/fm/article/view/3316/2764.

Habermas, Jürgen. 1981. *The Theory of Communicative Action*. (Vol. I). Boston: Beacon.

Habermas, Jürgen. 1987. *The Theory of Communicative Action*. (Vol. II). Boston: Beacon.

Habermas, Jürgen. 1990. 'Discourse Ethics: Notes on a Program of Philosophical Justification.' In Idem. *Moral Consciousness and Communicative Action*, 43–115. Cambridge, Mass: MIT Press.

Habermas, Jürgen. 1992. *Autonomy and Solidarity: Interviews with Jürgen Habermas*. Edited and introduced by Peter Dews. London, New York: Verso.

Habermas, Jürgen. 1996. *Between Facts and Norms: Contributions to a Discourse Theory of Law and Democracy*. London: Polity.

Habermas, Jürgen. 2001 [1993]. *Justification and Application: Remarks on Discourse Ethics*. Cambridge, Mass: MIT Press.

Haendel, Melissa A., Nicole A. Vasilevsky, and Jacqueline A. Wirz. 2012. 'Dealing with Data: A Case Study on Information and Data Management Literacy.' *PLoS Biology*, 10(5). Retrieved from: http://journals.plos.org/plosbiology/article?id=10.1371/journal.pbio.1001339.

Hahmann, Stefan, and Dirk Burghardt. 2013. 'How Much Information is Geospatially Referenced?' *Networks and Cognition. International Journal of Geographical Information Science*, 27(6): 1171–1189.

Haklay, Mordechai. 2013. 'Neogeography and the Delusion of Democratisation.' *Environment and Planning A: Economy and Space*, 45: 55–69.

Haklay, Mordechai, and Patrick Weber. 2008.' Openstreetmap: User-Generated Street Maps.' *Pervasive Computing, IEEE*, 7(4), 12–18.

Hampton, Stephanie E. et al. 2013. 'Big Data and the Future of Ecology.' *Frontiers in Ecology and the Environment*, 11(3): 156–162.

Hanssen, Beatrice. 2014. *Critique of Violence: Between Poststructuralism and Critical Theory*. Abingdon: Routledge.

Haraway, Donna J. 1997. *Modest_Witness@ Second-_Millennium. FemaleMan_ Meets_ OncoMouse: Feminism and Technoscience*. London: Routledge.

Haraway, Donna J. 1988. Situated Knowledges: The Science Question in Feminism and the Privilege of Partial Perspective. *Feminist Studies*, 14(3), 575–599.

Harding, Sandra G. 1986. *The Science Question in Feminism*. New York: Cornell University Press.

Harding, Sandra G. 2004, ed. *The Feminist Standpoint Theory Reader: Intellectual and Political Controversies*. London/New York: Routledge.

Harris, Anna, Sally Wyatt, and Susan Kelly. 2013a. 'The Gift of Spit (And the Obligation to Return it) How Consumers of Online Genetic Testing Services Participate in Research.' *Information, Communication & Society*, 16(2): 236–257.

Harris, Anna, Susan Kelly, and Sally Wyatt. 2013b. 'Counseling Customers: Emerging Roles for Genetic Counselors in the Direct-to-Consumer Genetic Testing Market.' *Journal of Genetic Counseling*, 22(2): 277–288.

Harris, Anna, Susan Kelly, and Sally Wyatt. 2016. *CyberGenetics: Health Genetics and New Media*. Abingdon: Routledge.

Hay, Simon et al. 2013. 'Big Data Opportunities for Global Infectious Disease Surveillance'. *PLoS Med*, 10(4): http://journals.plos.org/plosmedicine/article?id=10.1371/journal.pmed.1001413.

Heath, Alex. 2016. 'Mark Zuckerberg and his wife Priscilla Chan will invest $3 billion into curing all diseases by the end of this century'. *Business Insider*, September 21'. Retrieved from: http://uk.businessinsider.com/mark-zuckerberg-and-his-wife-will-invest-3-billion-into-curing-diseases-2016-9?r=US&IR=T

Hedgecoe, Adam. 2016. 'Scandals, Ethics, and Regulatory Change in Biomedical Research'. *Science, Technology, & Human Values*, 42(4): 577–599.

Heeks, Richard, and Jaco Renken. 2016. 'Data Justice for Development: What Would it Mean?' *Information Development*, online first, DOI: https://doi.org/10.1177/0266666916678282.

Heilig, Steve. 2015. 'Please don't change San Francisco General's name'. *San Francisco Chronicle*, April 2. Retrieved from: http://www.sfchronicle.com/opinion/openforum/article/Please-don-t-change-San-Francisco-General-s-6176553.php.

Helles, Rasmus and Klaus Jensen 2013. 'Making data – big data and beyond: Introduction to the special issue'. *First Monday*, 18(10). Retrieved from: http://firstmonday.org/ojs/index.php/fm/article/view/4860/3748.

Hermida, Alfred. 2012. 'Tweets And Truth: Journalism as a Discipline of Collaborative Verification'. *Journalism Practice*, 6(5-6):659–668.

Herzog, David. 2015. *Data Literacy: A User's Guide*. London: Sage.

Hicks, Diana. 2012. 'Performance-Based University Research Funding Systems'. *Research Policy*, 41(2): 251–261.

Hoedemaekers, Rogeer, Bert Gordijn, and Martien Pijnenburg. 2006. 'Does an Appeal to the Common Good Justify Individual Sacrifices for Genomic Research?' *Theoretical Medicine and Bioethics*, 27(5): 415–431.

Hoffman, David. 2014. 'Privacy Is a Business Opportunity'. *Harvard Business Review*. Retrieved from: https://hbr.org/2014/04/privacy-is-a-business-opportunity.

Hookway, Christopher. 2008. 'American Pragmatism: Fallibilism and Cognitive Progress'. In *Epistemology: The Key Thinkers*, edited by Stephen Hetherington, pp. 153–171. New York: Continuum.

Horvath, Aaron, and Walter W. Powell. 2016. 'Contributory or Disruptive: Do New Forms of Philanthropy Erode Democracy?' In Rob Reich, Chiara Cordelli, and Lucy Bernholz, eds., *Philanthropy in Democratic Societies: History, Institutions, Values*, pp. 87–112. Chicago: University of Chicago Press.

Hourihan, Matt and David Parkes. 2016. 'Federal R&D Budget Trends: A Short Summary'. *American Association for the Advancement of Science, report*. Retrieved from: https://mcmprodaaas.s3.amazonaws.com/s3fs-public/AAAS%20R%26D%20Budget%20Primer.pdf.

Howe, Doug et al. 2008. 'Big Data: The Future of Biocuration'. *Nature*, 455(7209): 47–50.

Iliadis, Andrew, and Federica Russo. 2016. 'Critical Data Studies: An Introduction'. *Big Data & Society*, 3(2): 1–7.

'Information and Communication Technologies in Horizon.' 2015. *European Commission: DG Connect.* Retrieved from: https://ec.europa.eu/digital-single-market/en/information-communication-technologies-horizon-2020.

Inverso, Gino, Nada Boualam, and Kevin B. Mahoney. 2017. 'The New Age of Private Research Funding: Be Careful Out There!' *Healthcare Transformation,* 2(2): 52–56.

Ioannidis, John. 2013. 'Informed Consent, Big Data, and the Oxymoron of Research That is Not Research.' *The American Journal of Bioethics,* 13(4): 40–42.

Ippolita. 2013. *The Dark Side of Google.* (Theory on Demand, Vol. 13). Amsterdam: Institute of Network Cultures.

Ireland, Molly E. et al. 'Action Tweets Linked to Reduced County-Level HIV Prevalence in the United States: Online messages and Structural Determinants.' *AIDS and Behavior,* 20(6): 1256–1264.

'IU School of Nursing and ChaCha partner to conduct interdisciplinary Big-Data research about health and wellness.' 2015. *IUPUI Newsroom,* February 2. Retrieved from: http://news.iupui.edu/releases/2015/02/News%20 Release.shtml.

Jackson, Jasper. 2017. 'Eli Pariser: Activist Whose Filter Bubble Warnings Presaged Trump and Brexit.' *The Guardian,* January 8. Retrieved from: https://www.theguardian.com/media/2017/jan/08/eli-pariser-activist-whose-filter-bubble-warnings-presaged-trump-and-brexit.

Jagadish, Hosagrahar et al. 2014. 'Big Data and its Technical Challenges.' *Communications of the ACM,* 57(7), 86–94.

Jardine, Jennifer et al. (2015). 'Apple's ResearchKit: Smart data collection for the smartphone era?' *Journal of the Royal Society of Medicine,* 108(8): 294-296.

Joas, Hans. 1993. *Pragmatism and Social Theory.* Chicago: University of Chicago Press.

Johnson, Jeffrey Alan. 2014. 'From Open Data to Information Justice.' *Ethics and Information Technology,* 16(4): 263–274.

Kahn, Jeffrey P., Effy Vayena, and Anna C. Mastroianni. 2014. 'Opinion: Learning as we go: Lessons from the Publication of Facebook's Social-Computing Research.' *Proceedings of the National Academy of Sciences,* 111(38), 13677–13679.

Kaplan, Frédéric. 2015. 'A Map for Big Data Research in Digital Humanities.' *Frontiers in Digital Humanities.* Retrieved from http://journal.frontiersin.org/article/10.3389/fdigh.2015.00001/full.

Kaplan, Robert S., and Michael E. Porter. 2011. How to Solve the Cost Crisis in Health Care. *Harvard Business Review,* 89(9): 46–52.

Kapp, Julie M., Colleen Peters, and Debra Parker Oliver. 2013. 'Research Recruitment Using Facebook Advertising: Big Potential, Big Challenges.' *Journal of Cancer Education,* 28(1): 134–137.

Keeler, Mark R. 2006. *Nothing to Hide: Privacy in the 21st Century.* New York: iUniverse.

Kelch, Robert P. 2002. 'Maintaining the Public Trust in Clinical Research.' *The New England Journal of Medicine*, 346(4): 285–287.

Keller, Mikaela et al. (2009). 'Use of Unstructured Event-Based Reports for Global Infectious Disease Surveillance.' *Emerging Infectious Diseases*, 15(5): 689–695.

Kennedy, Helen and Giles Moss. 2015. 'Known or Knowing Publics? Social Media Data Mining and the Question of Public Agency'. *Big Data & Society*, 2(2): 1–11.

Keulartz, Jozef, et al. 2002. *Pragmatist Ethics for a Technological Culture*. Dordrecht: Kluwer.

Keulartz, Jozef, et al. 2004. 'Ethics in Technological Culture: A Programmatic Proposal for a Pragmatist Approach.' *Science, Technology & Human Values*, 29(1): 3–29.

Kettl, Donald. 2017. *Little Bites of Big Data for Public Policy*. Washington: CQ Press.

Kim, Gang-Hoon, Silvana Trimi, and Ji-Hyong Chung. 2014. 'Big-Data Applications in the Government Sector.' *Communications of the ACM*, 57(3), 78–85.

Kirchner, Lauren. 2011. 'AOL Settled with Unpaid 'Volunteers' for $15 Million.' *Columbia Journalism Review*. Retrieved from: http://archives.cjr.org/the_news_frontier/aol_settled_with_unpaid_volunt.php.

Kirkpatrick, Robert. 2011. 'Data Philanthropy. Public & Private Sector Data Sharing for Global Resilience.' *United Nations Global Pulse Blog*. Retrieved from: http://www.unglobalpulse.org/blog/data-philanthropy-public-private-sector-data-sharing-global-resilience.

Kirkpatrick, Robert. 2016. 'The Importance of Big Data Partnerships for Sustainable Development.' *UN Global Pulse Blog*, May 31. Retrieved from: http://www.unglobalpulse.org/big-data-partnerships-for-sustainable-development.

Kitchin, Rob and Tracey P. Lauriault. 2014. Towards Critical Data Studies: Charting and Unpacking Data Assemblages and their Work. Retrieved from: http://papers.ssrn.com/sol3/papers.cfm?abstract_id=2474112.

Kitchin, Rob 2014a. *The Data Revolution: Big Data, Open Data, Data Infrastructures and Their Consequences*. London: Sage.

Kitchin, Rob 2014b. 'Big data, New Epistemologies and Paradigm Shifts'. *Big Data & Society*, 1(1), Advance online publication. DOI: https://doi.org/2053951714528481.

Kitchin, Rob, and Martin Dodge. 2011. *Code/Space: Software and Everyday Life*. Cambridge, Mass: MIT Press.

Kitchin, Rob. 'Big Data and Human Geography: Opportunities, Challenges and Risks.' *Dialogues in Human Geography*, 3(3): 262–267.

Kitchin, Rob. 2017. 'Thinking Critically about and Researching Algorithms.' *Information, Communication and Society*, 20(1): 14–29

Kittler, Friedrich. 1995. 'There is no software.' *Ctheory*. Retrieved from: https://journals.uvic.ca/index.php/ctheory/article/view/14655/5522

Klein, Hans K., and Daniel Lee Kleinman. 2002. 'The Social Construction of Technology: Structural Considerations.' *Science, Technology & Human Values*, 27(1): 28–52.

Kleinsman, John, and Sue Buckley. 2015. 'Facebook Study: A Little Bit Unethical But Worth It?' *Journal of Bioethical Inquiry*, 12(2): 179–182.

Kosinski, Michal, David Stillwell, and Thore Graepel. 2013. 'Private Traits and Attributes are Predictable from Digital Records of Human Behavior.' *Proceedings of the National Academy of Sciences*, 110(15): 5802–5805.

Kramer, Adam, Jamie E. Guillory, and Jeffrey T. Hancock. 2014. 'Experimental Evidence of Massive-Scale Emotional Contagion through Social Networks.' *Proceedings of the National Academy of Sciences*, 111(24): 8788–8790.

Kraska, Tim. 2013. 'Finding the Needle in the Big Data Systems Haystack.' *IEEE internet Computing*, (1), 84–86.

Kroes, Neelie. 2012. 'Digital Agenda and Open Data: From Crisis of Trust to Open Governing.' Retrieved from: http://europa.eu/rapid/press-release_SPEECH-12-149_en.htm.

Kshetri, Door Nir. 2016. *Big Data's Big Potential in Developing Economies: Impact on Agriculture, Health, and Environmental Security*. Boston, MA: Cabi.

LaFollette, Hugh. 2000. 'Pragmatic Ethics.' In Hugh LaFollette, eds., *The Blackwell Guide to Ethical Theory*, pp. 400–419. Oxford: Blackwell.

LaFrance, Adrienne. 2014. 'Even the Editor of Facebook's Mood Study Thought It Was Creepy.' *The Atlantic*, June 28. Retrieved from: http://www.theatlantic.com/technology/archive/2014/06/even-the-editor-of-facebooks-mood-study-thought-it-was-creepy/373649.

Lane, Julia et al., eds. 2014. *Privacy, Big Data, and the Public Good: Frameworks for Engagement*. New York: Cambridge University Press.

Latour, Bruno. 1999. *Pandora's Hope: Essays on the Reality of Science Studies*. Cambridge: Harvard University Press.

Law, John 2008. 'On Sociology and STS.' *The Sociological Review*, 56(4): 623–649.

Lazer, David and Ryan Kennedy. 2015. 'What Can we Learn from the Epic Failure of Google Flu Trends?' *Wired*. Retrieved from: http://www.wired.com/2015/10/can-learn-epic-failure-google-flu-trends.

Lazer, David et al. 2014. 'The Parable of Google Flu: Traps in Big Data Analysis.' *Science*, 343(6176): 1203–1205.

Leetaru, Kalev. 2016. 'Are Research Ethics Obsolete In The Era Of Big Data?' *Forbes*, June 17. Retrieved from: https://www.forbes.com/sites/kalev-leetaru/2016/06/17/are-research-ethics-obsolete-in-the-era-of-big-data/#3de593cd7aa3.

Levi, Michael, and David S. Wall. 2004. 'Technologies, Security, and Privacy in the Post-9/11 European Information Society.' *Journal of Law and Society*, 31(2), 194–220.

Levy, Karen EC, and David Merritt Johns. 2016. 'When Open Data is a Trojan Horse: The Weaponization of Transparency in Science and Governance.' *Big Data & Society*, 3(1): 1–6.

London, Alex J. 2003. 'Threats to the Common Good: Biochemical Weapons and Human Subjects Research.' *The Hastings Center Report*, 33(3): 17–25.

Lundh, Andreas et al. 2017. 'Industry Sponsorship and Research Outcome,' *Cochrane Database of Systematic Reviews*. Retrieved from: http://online library.wiley.com/doi/10.1002/14651858.MR000033.pub3/full.

Lupton, Deborah. 2013. 'Quantifying the Body: Monitoring and Measuring Health in the Age of mHealth Technologies'. *Critical Public Health*, 23(4), 393–403.

Lupton, Deborah. 2014. 'You Are Your Data: Self-Tracking Practices and Concepts of Data.' Retrieved from: http://papers.ssrn.com/sol3/Papers.cfm?abstract_id=2534211.

Lupton, Deborah. 2014a. Beyond Techno-Utopia: Critical Approaches to Digital Health Technologies. *Societies*, 4(4), 706–711.

Lupton, Deborah. 2014b. *Digital Sociology*. Abingdon: Routledge.

Lupton, Deborah. 2014c. Critical Perspectives on Digital Health Technologies. *Sociological Compass*, 8(12), 1344–1359.

Lupton, Deborah. 2014d. 'The Commodification of Patient Opinion: The Digital Patient Experience Economy in the Age of Big Data.' *Sociology of Health & Illness*, 36(6): 856–869.

Lupton, Deborah. 2015. 'Quantified Sex: A Critical Analysis of Sexual and Reproductive Self-Tracking Using Apps.' *Culture, Health & Sexuality*, 17(4): 440–453.

Lupton, Deborah. 2015. The Thirteen Ps of Big Data. *This Sociological Life*. Retrieved from: https://www.researchgate.net/profile/Deborah_Lupton/publication/276207564_The_Thirteen_Ps_of_Big_Data/links/5552c2d808ae6fd2d81d5f20.pdf.

Lupton, Deborah. 2016. 'You Are Your Data: Self-Tracking Practices and Concepts of Data.' In: Stefan Selke, ed., *Lifelogging*, pp. 61–79. Wiesbaden: Springer.

Luscombe, Nicholas, Dov Greenbaum, and Mark Gerstein. 2001. 'What is Bioinformatics? A Proposed Definition and Overview of the Field.' *Methods of Information in Medicine*, 40(4): 346–358.

Lyon, Aidan et al. 2012. 'Comparison of Web-Based Biosecurity Intelligence Systems: BioCaster, EpiSPIDER and HealthMap'. *Transboundary and emerging diseases*, 59(3): 223–232.

Lyon, David. 2014. 'Surveillance, Snowden, and Big Data: Capacities, Consequences, Critique.' *Big Data & Society*, 1(2): DOI: https://doi.org/2053951714541861.

Lysaught, M. Therese. 2004. 'Respect: or, How Respect for Persons Became Respect for Autonomy.' *Journal of Medicine and Philosophy*, 29(6): 665–680.

Malin, Bradley A., Khaled El Emam, and Christine M. O'Keefe. 2013. 'Biomedical Data Privacy: Problems, Perspectives, and Recent Advances.' *Journal of the American Medical Informatics Association*, 20(1): 2–6.

Manovich, Lev. 2011. 'Trending: The Promises and the Challenges of Big Social Data'. Retrieved from: https://pdfs.semanticscholar.org/15ff/fafb4bfcf8f9b210de01ac5208b0d916147e.pdf.

Manovich, Lev. 2013. *Software Takes Command*. New York: Bloomsbury.

Marchant, Gary E., Braden R. Allenby, and Joseph R. Herkert, eds. 2011. *The Growing Gap Between Emerging Technologies and Legal-Ethical Oversight: The Pacing Problem*. London/New York: Springer.

Margolis, Ronald et al. 2014. 'The National Institutes of Health's Big Data to Knowledge (BD2K) Initiative: Capitalizing on Biomedical Big Data'. *Journal of the American Medical Informatics Association*, 21(6): 957–958.

Margonelli, Liza. 1999. 'Inside AOL's 'Cyber-Sweatshop'', *Wired*, January 10, Retrieved from: https://www.wired.com/1999/10/volunteers.

'Mark Zuckerberg and Priscilla Chan give $75 million to support San Francisco General Hospital and Trauma Center'. 2015. *San Francisco General Hospital Foundation*, February 6. Retrieved from: https://sfghf.org/rebuild/mark-zuckerberg-priscilla-chan-give-75-million-support-san-francisco-general-hospital-trauma-center.

Marr, Bernard. 2015. *Big Data: Using SMART Big Data, Analytics and Metrics to Make Better Decisions and Improve Performance*. Hobeken, NJ: John Wiley & Sons.

Marz, Nathan, and James Warren. 2012. *Big Data: Principles and Best Practices of Scalable Realtime Data Systems*. Westhampton: Manning.

Mattioli, Michael. 2014. 'Disclosing Big Data'. *Minnesota Law Review*, 99, 535–583.

Mattmann, Chris A. 2013. 'Computing: A Vision for Data Science'. *Nature*, 493(7433), 473–475.

Mayer-Schönberger, Viktor, and Kenneth Cukier. 2013. *Big Data: A Revolution That Will Transform How We Live, Work, and Think*. Boston, New York: Houghton Mifflin Harcourt.

McCarthy, Thomas. 2007. 'Introduction'. In *Moral Consciousness and Communicative Action*, by Jürgen Habermas, pp. vi–xii. London: Polity.

McCullagh, Declan. 2008. 'Privacy Groups Target Google Flu Trends'. Retrieved from: https://www.cnet.com/news/privacy-groups-target-google-flu-trends.

McFall, Liz. 2015: Is Digital Disruption the End of Health Insurance? Some Thoughts on the Devising of Risk. *Econstor*, 17(1): 32–44. Retrieved from: http://hdl.handle.net/10419/156065.

Meier, Patrick. 2014. 'Crisis Mapping Haiti: Some Final Reflections'. Retrieved from: http://reliefweb.int/report/haiti/crisis-mapping-haiti-some-final-reflections.

Meier, Patrick. 2015. *Digital Humanitarians: How Big Data is Changing the Face of Humanitarian Response*. Boca Raton/London/New York: CRC.

Mercom. 2016. 'Healthcare IT'. Retrieved from: http://mercomcapital.com/healthcare-it-has-strong-first-quarter-with-$1.4-billion-in-vc-funding-reports-mercom-capital-group.

Mearian, Lucas. 2015. Insurance company now offers discounts – if you let it track your Fitbit. *Computer World*, July 18. Retrieved from: https://www.computerworld.com/article/2911594/insurance-company-now-offers-discounts-if-you-let-it-track-your-fitbit.html.

Metcalf, Jacob and Crawford, Kate. 2016. Where are Human Subjects in Big Data Research? The Emerging Ethics Divide. *Big Data & Society*, 3(1), DOI: https://doi.org/20539517166650211.

Milan, Stefania. 2016. 'Data Activism: The Politics of Big Data According to Civil Society.' Available at: https://data-activism.net/about.

Mingers, John, and Geoff Walsham. 2010. 'Toward Ethical Information Systems: The Contribution of Discourse Ethics.' *MIS Quarterly*, 34(4): 833–854.

Mittelstadt, Brent. 2013. *On the Ethical Implications of Personal Health Monitoring: A Conceptual Framework for Emerging Discourses*. Leicester: De Montfort University.

Mittelstadt, Brent and Floridi, Luciano 2016@[2015 online first]. 'The Ethics of Big Data: Current and Foreseeable Issues in Biomedical Contexts'. *Science and Engineering Ethics*, 22: 303–341.

Mittelstadt, Brent and Floridi, Luciano, eds. 2016. *The Ethics of Biomedical Big Data*. London: Springer.

Mittelstadt, Brent D., Bernd C. Stahl and Ben Fairweather. 2015. 'How to Shape a Better Future? Epistemic Difficulties for Ethical Assessment and Anticipatory Governance of Emerging Technologies.' *Ethical Theory and Moral Practice*, 1–21. Dordrecht: Springer.

Moor, James H. 2005. Why We Need Better Ethics for Emerging Technologies. *Ethics and Information Technology*, 7(3): 111–119.

Morrissey, Simon. 2016. 'Take Notice! The Legal and Commercial Impact of the General Data Protection Regulation's Rules on Privacy Notices.' *Journal of Data Protection & Privacy* 1(1): 46–52.

Mosco, Vincent. 2015. *To the Cloud: Big Data in a Turbulent World*. Boulder, London: Paradigm.

Moser, Susanne C. 2014. 'Communicating Adaptation to Climate Change: The Art and Science of Public Engagement When Climate Change Comes Home.' *Wiley Interdisciplinary Reviews: Climate Change*, 5(3): 337–358.

Mulder, Femke, Julie Ferguson, Peter Groenewegen, Kees Boersma, and Jeroen Wolbers. 2016. 'Questioning Big Data: Crowdsourcing Crisis Data Towards an Inclusive Humanitarian Response.' *Big Data & Society*, 3(2), DOI: https://doi.org/20539517166662054.

Nambisan, Priya et al. 2015. 'Social Media, Big Data, and Public Health Informatics: Ruminating Behavior of Depression Revealed Through Twitter.' *System Sciences. 48th Hawaii International Conference*: 2906–2913.

Nature Editorial. 2007. 'A Matter of Trust.' Retrieved from: http://www.nature.com/nature/journal/v449/n7163/full/449637b.html.

Naur, Peter. 1966. 'The Science of Datalogy.' *Communications of the ACM*, 9(7): 485.

Nederbragt, Hubertus. 2000. 'The Biomedical Disciplines and the Structure of Biomedical and Clinical Knowledge.' *Theoretical Medicine and Bioethics*, 21(6): 553–566.

Neff, Gina et al. 2017. 'Critique and Contribute: A Practice-Based Framework for Improving Critical Data Studies and Data Science.' *Big Data*, 5(2): 85–97.

Newman, Abraham L. (2015). 'What the 'Right to be Forgotten' Means for Privacy in a Digital Age.' *Science*, 347(6221): 507–508.

Niu, Evan. 2017. 'Twitter, Inc. Can Survive.' Nasday, February 13. Retrieved from: http://www.nasdaq.com/article/twitter-inc-can-survive-cm747325.

'Not Using Data is the Moral Equivalent of Burning Books'. 2016. *Digitising Europe Initiative Brussels 2016, Vodafone Institute for Society and Communications*. Retrieved from: http://www.vodafone-institut.de/event/not-using-data-is-the-moral-equivalent-of-burning-books.

Nyrén, Olof, Magnus Stenbeck, and Henrik Grönberg. 2014. 'The European Parliament Proposal for the New EU General Data Protection Regulation May Severely Restrict European Epidemiological Research.' *European Journal of Epidemiology*, 29(4): 227–230.

O'Driscoll, Aisling, Jurate Daugelaite, and Roy D. Sleator. 2013. 'Big Data, Hadoop and Cloud Computing in Genomics.' *Journal of Biomedical Informatics*, 46(5): 774–781.

O'Hara, Kieron. 2011. 'Transparent Government, Not Transparent Citizens: A Report on Privacy and Transparency for the Cabinet Office.' *Cabinet Office London*. Retrieved from: https://www.gov.uk/government/publications/independent-transparency-and-privacy-review.

Oboler, Andre, Kristopher Welsh, and Lito Cruz. 2012. 'The Danger of Big Data: Social Media as Computational Social Science.' *First Monday* 17(7). Retrieved from: http://firstmonday.org/ojs/index.php/fm/article/viewArticle/3993.

Ohlhorst, Frank 2012. *Big Data Analytics: Turning Big Data into Big Money*. Hobeken, NJ: John Wiley & Sons.

Ohm, Paul. 2010. 'Broken Promises of Privacy: Responding to the Surprising Failure of Anonymization.' *UCLA Law Review*, 57: 1701–1777.

Open Knowledge International. n.d. 'What is Open Data?' *Open Data Handbook*. Retrieved from: http://opendatahandbook.org/guide/en/what-is-open-data.

Orton-Johnson, Kate, and Nick Prior, eds. 2013. *Digital Sociology: Critical Perspectives*. Houndmills, Basingstoke: Palgrave.

Oser, Carrie-Beth. n.d. 'Social Networks and HIV Risk Behaviors of Special Population Drug Users'. Retrieved from: https://projectreporter.nih.gov/project_info_description.cfm?aid=9088432&icde=31609271.

Ostfeld, Richard S. et al. (2005). 'Spatial Epidemiology: An Emerging (or Re-Emerging) Discipline'. *Trends in Ecology & Evolution*, 20(6): 328–336.

Oudshoorn, Nelly and Trevor Pinch, eds. 2003. *How Users Matter: The Co-Construction of Users and Technology*. Cambridge, Mass: MIT.

Oulasvirta, Antti et al. 2014. 'Transparency of Intentions Decreases Privacy Concerns in Ubiquitous Surveillance'. *Cyberpsychology, Behavior, and Social Networking*, 17(10): 633–638.

Parikka, Jussi. 2015. *A Geology of Media*. Minneapolis, London: University of Minnesota Press.

Pariser, Eli. 2011. *The Filter Bubble: What the Internet is Hiding from You*. New York: Penguin.

Pariser, Eli. 2015. 'Did Facebook's Big New Study Kill My Filter Bubble Thesis?' Retrieved from: https://backchannel.com/facebook-published-a-big-new-study-on-the-filter-bubble-here-s-what-it-says-ef31a292da95#.6afqkrq7z.

Park, Sungmee and Sundaresan Jayaraman (2014, August). A transdisciplinary approach to wearables, big data and quality of life. In: *Engineering in Medicine and Biology Society (EMBC), 36th Annual International Conference of the IEEE*, pp. 4155–4158.

Parry, Bronwyn, and Beth Greenhough. 2018. *Bioinformation*. Cambridge: Polity.

Parsons, Michael D. 2004. 'Lobbying in Higher Education'. In. Edward John & Michael Parsons, eds., *Public Funding of Higher Education: Changing Contexts and New Rationales*, pp. 215–230, Baltimore/London: Johns Hopkins.

Paul, Michael et al. 2016. 'Social Media Mining for Public Health Monitoring and Surveillance'. *Pacific Symposium on Biocomputing*: Retrieved from https://psb.stanford.edu/psb-online/proceedings/psb16/intro-smm.pdf.

Paul, Michael and Mark Dredze. 2017. *Social Monitoring for Public Health*. San Rafael, CA: Morgan & Claypool.

Pawelke, Andreas and Anoush Rima Tatevossian. 2013. 'Data Philanthropy. Where are we Now?' *United Nations Global Pulse Blog*. Retrieved from: http://www.unglobalpulse.org/data-philanthropy-where-are-we-now.

Pinch, Trevor. 1996. 'The Social Construction of Technology: A review'. In Robert Fox ed., *Technological Change: Methods and Themes in the History of Technology*, , pp. 17–35. Australia: Harwood Academic Publishers.

Points Group. n.d. 'Investing in Google Ads or Facebook Ads for Healthcare'. Retrieved from: https://www.pointsgroupllc.com/investing-google-ads-facebook-ads-healthcare.

Polgreen, Philip M. et al. 2008. 'Using internet searches for influenza surveillance'. *Clinical Infectious Diseases*, 47(11): 1443–1448.

Pries, K. H., and Dunnigan, R. 2015. *Big data Analytics: A practical Guide for Managers*. London: CRC Press.

Puschmann, Cornelius and Burgess, Jean. 2013. 'The Politics of Twitter Data'. *HIIG Discussion Paper Series*. Retrieved from: http://tinyurl.com/j2eerbg.

Pybus, Jennifer, Mark Coté, and Tobias Blanke. 'Hacking the Social Life of Big Data.' *Big Data & Society* , 2(2): 1–10, DOI: https://doi.org/10.1177/2053951715616649.

Radder, Hans. 1998. 'The Politics of STS.' *Social Studies of Science* 28(2): 325–331.

Rana, Sanjay and Thierry Joliveua. 2009. 'NeoGeography: An Extension of Mainstream Geography for Everyone Made by Everyone?' *Journal of Location Based Services*, 3(2): 75-81.

Rathenau Institute. 2016. 'Public Trust in Science.' Retrieved from: https://www.rathenau.nl/en/page/public-trust-science.

Rehg, William. 1994. *Insight and Solidarity: The Dscourse Ethics of Jürgen Habermas*. Berkeley/Los Angeles/London: University of California Press.

Rehg, William. 2015. 'Discourse Ethics for Computer Ethics: A Heuristic for Engaged Dialogical Reflection.' *Ethics and Information Technology*, 17(1): 27–39.

Reich, Rob, Chiara Cordelli, and Lucy Bernholz, eds. 2016 *Philanthropy in Democratic Societies: History, Institutions, Values.* Chicago: University of Chicago Press.

Reilly, Michael. 2008. 'Google the Next Emerging Pandemic With HealthMap.' Retrievedfrom:http://io9.gizmodo.com/5024280/google-the-next-emerging-pandemic-with-healthmap.

Research Councils UK n.d. 'Big data.' Retrieved from: http://www.rcuk.ac.uk/research/infrastructure/big-data.

'ResearchKit and CareKit'. (n.d.). Retrieved from https://www.apple.com/lae/researchkit.

Ribes, David, and Steven J. Jackson. 2013. 'Data Bite Man: The Work of Sustaining a Long–Term Study.' In: Lisa Gitelman, ed., *Raw Data is an Oxymoron*, pp. 147–166. Cambridge, Massachusetts: MIT Press.

Richards, Evelleen, and Malcolm Ashmore. 1996. '"More Sauce Please!' The Politics of SSK: Neutrality, Commitment and Beyond.' *Social Studies of Science*: 1: 219–228.

Richterich, Annika. 2014a. Google Trends: Using and Promoting Search Volume Indicators for Research. *Mediale Kontrolle unter Beobachtung*, 3(1). Retrieved from: www.medialekontrolle.de/wp-content/uploads/2014/09/Richterich-Annika-2014-03-01.pdf.

Richterich, Annika. 2014b. Von 'Supply' zu 'Demand' Google Flu Trends und transaktionale big data in der epidemiologischen Surveillance. In: Reichert, R. (Ed.). (2014). *Big data: Analysen zum digitalen Wandel von Wissen, Macht und Ökonomie*. Bielefeld: transcript.

Richterich, Annika. 2016. 'Google Flu Trends and Ethical Implications of 'Info-demiology'. In: Brent Mittelstadt and Luciano Floridi, eds., *The Ethics of Biomedical Big Data*, pp. 41–72. London: Springer.

Richterich, Annika. 2018 'Digital Health Mapping: Big Data Utilisation and User Involvement in Public Health Surveillance.' In: T. Felgenhauer and K. Gäbler, eds., *Geographies of Digital Culture*, Abingdon: Routledge.

Rieder, Bernhard. (2010). One network and four algorithms. The Politics of Systems. Retrieved from: http://thepoliticsofsystems.net/2010/10/one-network-and-four-algorithms.

Rieder, Bernhard. 2016, May 27. Closing APIs and the public scrutiny of very large online platforms. Retrieved from: http://thepoliticsofsystems.net/2016/05/closing-apis-and-the-public-scrutiny-of-very-large-online-platforms.

Rieder, Gernot, and Judith Simon. 2016. 'Datatrust: Or, the Political Quest for Numerical Evidence and the Epistemologies of Big Data.' *Big Data & Society* 3(1): DOI: https://doi.org/2053951716649398.

Rip, Arie. 2013. 'Pervasive Normativity and Emerging Technologies.' In Simone van der Burg and Tsjalling Swierstra, eds., *Ethics on the Laboratory Floor*, pp. 191–212. Basingstoke, Hampshire: Palgrave Macmillan.

Ritzer, George, and Nathan Jurgenson. 2010. 'Production, Consumption, Prosumption. The Nature of Capitalism in the Age of the Digital 'Prosumer'.' *Journal of Consumer Culture*, 10(1): 13–36.

Roberts, P. 2005. *The End of Oil: The Decline of the Petroleum Economy and the Rise of a New Energy Order*. London: Bloomsbury Publishing.

Rogers, Richard. 2013. *Digital Methods*. Cambridge MA: MIT press.

Rorty, Richard. 1992. 'What Can You Expect from Anti-Foundationalist Philosophers?: A Reply to Lynn Baker.' *Virginia Law Review*: 719–727.

Rorty, Richard. 1994. 'Inquiry as Recontextualization – An Anti-Dualist Account of Interpretation.' *Filosoficky Casopis*, 42 (3): 358–379.

Rothstein, Mark A. (2015). 'Ethical Issues in Big Data Health Research: Currents in Contemporary Bioethics.' *The Journal of Law, Medicine & Ethics*, 43(2), 425–429.

Rothstein, Mark A., and Abigail B. Shoben. 2013. 'Does Consent Bias Research?' *The American Journal of Bioethics*, 13(4): 27–37.

Rowe, Gene, and Lynn J. Frewer. 2005. 'A Typology of Public Engagement Mechanisms.' *Science, Technology, & Human Values*, 30(2): 251–290.

Rowland, Katherine. 2012. 'Epidemiologists Put Social Media in the spotlight.' *Nature*. Retrieved from: http://www.nature.com/news/epidemiologists-put-social-media-in-the-spotlight-1.10012.

Salganic, Matthew. 2017. *Bit by Bit: Social Research in the Digital Age*. Princeton, NJ: Princeton University Press.

SC1-PM-18. 2016. 'Big data supporting Public Health Policies.' Retrieved from: https://ec.europa.eu/eip/ageing/funding/horizon-2020/big-data-supporting-public-health-policies-sc1-pm-18-2016_en.

Schöch, Christof. 2013. 'Big? Smart? Clean? Messy? Data in the Humanities.' *Journal of Digital Humanities*, 2(3): 2–13.

Scholz, Trebor, ed. 2012. *Digital Labor: The Internet as Playground and Factory*. New York and London: Routledge.

Schrock, Andrew. 2016. 'Civic Hacking as Data Activism and Advocacy: A History from Publicity to Open Government Data.' *New Media & Society*, 18(4): 581–599.

Schroeder, Ralph. 2014. 'Big Data and the Brave New World of Social Media Research.' *Big Data & Society*, 1(2), DOI: https://doi.org/2053951714563194.

Science and Technology Committee (2015). 'The Big Data Dilemma'. Retrieved from: http://www.publications.parliament.uk/pa/cm201516/cmselect/cmsctech/468/46809.htm#_idTextAnchor035.

Selgelid, Michael J., Angela McLean, Nimalan Arinaminpathy, and Julian Savulescu, eds. 2011. *Infectious Disease Ethics: Limiting Liberty in Contexts of Contagion*. Netherlands: Springer.

Sharon, Tamar and Zandbergen, Dorien. 2016. 'From Data Fetishism to Quantifying Selves: Self-Tracking Practices and the Other Values of Data.' *New Media & Society*, advance online publication, DOI: https://doi.org/1461444816636090.

Sharon, Tamar. 2016. 'The Googlization of Health Research: From Disruptive Innovation to Disruptive Ethics.' *Personalized Medicine*. Retrieved from: http://www.futuremedicine.com/doi/abs/10.2217/pme-2016-0057.

Sharon, Tamar. 2016a. Apple and Google Plan to Reinvent Health Care. Should We Worry? *The Hastings Center: Bioethics Forum*. Retrieved from: http://www.thehastingscenter.org/apple-and-google-plan-to-reinvent-health-care-should-we-worry.

Shelton, Taylor et al. 2014. 'Mapping the Data Shadows of Hurricane Sandy: Uncovering the Sociospatial Dimensions of 'Big Data.' *Geoforum* 52: 167–179.

Signorini, Alessio et al. 2011. 'The Use of Twitter to Track Levels of Disease Activity and Public Concern in the US During the Influenza A H1N1 Pandemic.' *PloS one*, 6(5): http://journals.plos.org/plosone/article?id=10.1371/journal.pone.0019467.

Silicon Valley Bank. 2017. 'Trends in Healthcare Investments and Exits.' Retrieved from: https://www.svb.com/uploadedFiles/Content/Trends_and_Insights/Reports/Healthcare_Investments_and_Exits_Report/healthcare-report-2017.pdf.

Simon, Phil. 2013. *Too Big to Ignore: The Business Case for Big Data*. Hoboken, NJ: John Wiley & Sons.

Solove, Daniel J. 2011. *Nothing to Hide: The False Tradeoff Between Privacy and Security*. New Haven, Connecticut: Yale University Press.

Stilgoe, Jack, Simon J. Lock, and James Wilsdon. 2014. 'Why Should we Promote Public Engagement with Science?' *Public Understanding of Science*, 23(1): 4–15.

Stodden, Victoria. 2014. 'Enabling Reproducibility in Big Data Research: Balancing Confidentiality and Scientific Transparency.' In Julia Lane et al., eds., *Privacy, Big Data, and the Public Good: Frameworks for Engagement*, pp. 112–132, New York: Cambridge University Press.

Stoové, Mark A., and Alisa E. Pedrana. 2014. 'Making the Most of a Brave New World: Opportunities and Considerations for Using Twitter as a Public Health Monitoring Tool.' *Preventive Medicine*, 63: 109–111.

Sveinsdottir, Edda, and Erik Frøkjær. 1988. 'Datalogy – the Copenhagen Tradition of Computer Science.' *BIT Numerical Mathematics*, 28(3): 450–472.

Swierstra, Tsjalling and Rip, Arie. 2007. Nano-Ethics as NEST-Ethics: Patterns of Moral Argumentation About New and Emerging Science and Technology. *Nanoethics*, 1(1): 3–20.

Tapscott, Don. 1996. *The Digital Economy*. New York: McGraw-Hill.

Taylor, Linnet. 2015. 'No Place to Hide? The Ethics and Analytics of Tracking Mobility using Mobile Phone Data.' *Environment and Planning D: Society and Space*, Advance online publication. DOI: https://doi.org/0263775815608851.

Taylor, Linnet, Luciano Floridi and Bart van der Sloot, eds. 2016. *Group Privacy: New Challenges of Data Technologies*. London: Springer.

Taylor, Linnet. 2017. 'What Is Data Justice? The Case for Connecting Digital Rights and Freedoms Globally.' Available at: http://dx.doi.org/10.2139/ssrn.2918779.

Tene, Omer, and Jules Polonetsky. 2012. 'Big Data for All: Privacy and User Control in the Age of Analytics.' *Northwestern Journal of Technology and Intellectual Property*, 239: 243–251.

Tene, Omer, and Jules Polonetsky. 2012a. 'Privacy in the Age of Big Data: A Time for Big Decisions.' *Stanford Law Review Online* 64: 63.

Terranova, Tiziana. 2000. 'Free Labor: Producing Culture for the Digital Economy.' *Social text*, 18(2): 33–58.

The European Parliament and the Council of the European Union. 2016. 'General Data Protection Regulation.' *Official Journal of the European Union*, April 27. Retrieved from: http://ec.europa.eu/justice/data-protection/reform/files/regulation_oj_en.pdf.

'The Health Data Ecosystem and Big Data.' n.d. *World Health Organisation: eHealth*. Retrieved from: http://www.who.int/ehealth/resources/ecosystem/en.

Thorp, Jer. 2012. 'Big Data Is Not the New Oil.' *Harvard Business Review*. Retrieved from: https://hbr.org/2012/11/data-humans-and-the-new-oil.

Toffler, Alvin. 1970. *Future Shock*. New York: Amereon.

Tolentino, Herman et al. (2007). 'Scanning the Emerging Infectious Diseases Horizon-visualizing ProMED emails using EpiSPIDER.' *Advances in Disease Surveillance*, 2(169): https://lhncbc.nlm.nih.gov/files/archive/pub2007055.pdf.

Tonidandel, Scott, King, Eden B. and Cortina, Jose M. 2016. *Big Data at Work: The Data Science Revolution and Organizational Psychology*. London: Taylor & Francis.

UN Global Pulse. 2012. 'Big Data for Development: Challenges and Opportunities.' Retrieved from: http://www.unglobalpulse.org/sites/default/files/BigDataforDevelopment-UNGlobalPulseJune2012.pdf.

'United Nations Global Pulse: About'. n.d. Retrieved from: http://www.unglobalpulse.org/about-new.

Uprichard, Emma. 2013. 'Big data, Little Questions?'. Retrieved from: http://discoversociety.org/wp-content/uploads/2013/10/DS_Big-Data.pdf.

Uršič, Helena. 2016. 'The Right to be Forgotten or the Duty to be Remembered? Twitter Data Reuse and Implications for User Privacy.' *Council for Big Data,*

Ethics, and Society. Retrieved from: http://bdes.datasociety.net/wp-content/uploads/2016/10/Ursic-politiwoops.pdf.

Vaidhyanathan, Siva. 2012. *The Googlization of Everything (and Why We Should Worry)*. Berkeley, LA: University of California Press.

Van Dijck, José. 2011. 'Tracing Twitter: The Rise of a Microblogging Platform.' *International Journal of Media and Cultural Politics*, 7(3), 333–348.

Van Dijck, José. 2013. *The Culture of Connectivity: A Critical History of Social Media*. New York: Oxford University Press.

Van Dijck, José. 2014. 'Datafication, Dataism and Dataveillance: Big Data Between Scientific Paradigm and Ideology.' *Surveillance & Society* 12(2), 197–208.

Vayena, Effy et al. 2015. 'Ethical Challenges of Big Data in Public Health.' *PLoS Computational Biology*, 11(2). Retrieved from: http://dx.doi.org/10.1371/journal.pcbi.1003904.

Velasco, Edward et al. 2014. 'Social Media and internet-Based Data in Global Systems for Public Health Surveillance: A Systematic Review'. *Milbank Quarterly*, 92(1): 7–33.

Vincent, James. 2016. 'Snapchat applies for patent to serve ads by recognizing objects in your snaps.' *The Verge*. Retrieved from: www.theverge.com/2016/7/18/12211292/snapchat-object-recognition-advertising.

Vis, Farida. 2013. 'A Critical Reflection on Big Data: Considering APIs, Researchers and Tools as Data Makers', *First Monday*, 18(10). Retrieved from: http://firstmonday.org/ojs/index.php/fm/article/view/4878/3755.

Volz, Dustin and Supantha Mukherjee. 2016. 'Twitter cuts jobs with eye on 2017 profit' *Reuters*, Octover 27. Retrieved from: http://www.reuters.com/article/us-twitter-results-idUSKCN12R1GW.

Wajcman, Judy. 2007. 'From Women and Technology to Gendered Technoscience.' *Information, Community and Society* 10(3): 287–298.

Wang, Wei et al. n.d. 'Mining the social web to monitor public health and HIV risk behaviors.' Retrieved from: https.//projectreporter.nih.gov/project_info_description.cfm?aid=9146666&icde=31609271.

Wauters, Robin. 2009. 'ChaCha Co-Founder Brad Bostic Steps Down As President'. *TechCrunch*, April 30. Retrieved from: https://techcrunch.com/2009/04/30/chacha-co-founder-brad-bostic-steps-down-as-president.

'Wearing Welness on your Sleeve'. 2016. *Northwestern School of Professional Studies*, March 21. http://sps.northwestern.edu/main/news-stories/wearable-technology-healthcare.php.

Weaver, David A., and Bruce Bimber (2008). 'Finding News Stories: A Comparison of Searches Using LexisNexis and Google News'. *Journalism & Mass Communication Quarterly*, 85(3): 515–530.

Weber, Jutta. 2006. 'From Science and Technology to Feminist Technoscience.' *Handbook of Gender and Womens Studies*, edited by Kathy Davis, Mary Evans, and Judith Lorber, 397–414. London: Sage.

Weindling, Paul. 1993. 'Public Health and Political Stabilisation: The Rockefeller Foundation in Central and Eastern Europe between the Two World Wars.' *Minerva*, 31(3): 253–267.

Weindling, Paul. 2001. 'The Origins of Informed Consent: The International Scientific Commission on Medical War Crimes, and the Nuremberg Code.' *Bulletin of the History of Medicine*, 75(1): 37–71.

Welch, Kelly. 2007. 'Black Criminal Stereotypes and Racial Profiling.' *Journal of Contemporary Criminal Justice*, 23(3): 276–288.

Wessels, Bridgette. 2015. 'Authentication, Status, and Power in a Digitally Organized Society.' *International Journal of Communication*, 9: 2801–2818.

Westra, Anna et al. (2014). New EU Clinical Trials Regulation: Needs a Few Tweaks before Implementation. *The BMJ*, 348. Retrieved from: http://www.bmj.com/content/348/bmj.g3710.

Wilhelm, A. 2014. Facebook Reading Android Users' Texts? Well, Hold On. *TechCrunch*. Retrieved from: http://techcrunch.com/2014/01/28/facebook-reading-android-users-texts-well-hold-on.

Wilsdon, James, and Rebecca Willis. 2004. *See-Through Science: Why Public Engagement Needs to Move Upstream*. London: Demos.

Winner, Langdon. 1980. 'Do Artifacts Have Politics?' *Daedalus*, 109(1): 121–36.

Winner, Langdon. 1993. 'Upon Opening the Black Box and Finding it Empty: Social Constructivism and the Philosophy of Technology.' *Science, Technology, & Human Values*, 18(3): 362–78.

World Wide Web Foundation. 2016. 'Open Data Barometer, 3rd Edition: Global Report.' Retrieved from: http://opendatabarometer.org/3rdedition/report.

Wyatt, Sally, Anna Harris, Samantha Adams, and Susan E. Kelly 2013. 'Illness Online: Self-Reported Data and Questions of Trust in Medical and Social Research.' *Theory, Culture & Society*, 30(4): 131–150.

Wynne, Brian. 2006. 'Public Engagement as a Means of Restoring Public Trust in Science–Hitting the Notes, but Missing the Music?' *Public Health Genomics*, 9(3): 211–220.

Yarkoni, Tal. 2014. 'In defense of Facebook.' *Personal blog*, June 28. Retrieved from: http://www.talyarkoni.org/blog/2014/06/28/in-defense-of-facebook.

Yarkoni, Tal. 2014a. 'In defense of In Defense of Facebook.' *Personal blog*, July 1. Retrieved from: http://www.talyarkoni.org/blog/2014/07/01/in-defense-of-in-defense-of-facebook.

Young, Sean D., Caitlin Rivers, and Bryan Lewis. 2014. 'Methods of Using Real-time Social Media Technologies for Detection and Remote Monitoring of HIV Outcomes.' *Preventive Medicine*, 63: 112–115.

Young, Sean D., Wenchao Yu, and Wei Wang. 2017. 'Toward Automating HIV Identification: Machine Learning for Rapid Identification of HIV-Related Social Media Data.' *Journal of Acquired Immune Deficiency Syndromes*, 74: 128–S131.

Young, Sean. n.d. 'Mining real-time social media big data to monitor HIV: Development and ethical issues'. Retrieved from: https://projectreporter. nih.gov/project_info_description.cfm?aid=9317061&icde=31609271.

Zetter, Kim. 2006. 'Brilliant's Wish: Disease Alerts.' *Wired*, February 23. Retrieved from: http://archive.wired.com/science/discoveries/news/2006/02/70280?cu rrentPage=all.

Zhang, Yulei et al. (2009). 'Automatic Online News Monitoring and Classification for Syndromic Surveillance'. *Decision Support Systems*, 47(4): 508–517.

Zikopoulos, Paul et al. 2012. *Understanding Big Data*. New York: McGraw Hill.

Zimmer, Michael and Nicholas Proferes. 2014. 'Privacy on Twitter, Twitter on Privacy'. In: Katrin Weller et al., eds. *Twitter and Society*, pp. 169–183. New York: Peter Lang.

Zimmer, Michael. 2010. "But the Data is Already Public': On the Ethics of Research in Facebook.' *Ethics and Information Technology*, 12(4): 313–325.

Zwitter, Andrej. 2014. 'Big Data Ethics'. *Big Data & Society*, 1(2): http://journals. sagepub.com/doi/full/10.1177/2053951714559253.

Index

Printed and bound by CPI Group (UK) Ltd, Croydon, CR0 4YY

07/07/2026

14916227-0001